Tomorrow the Moon

The Story of the Space Age

Joseph F. Alcock

Aerospace Publishers

PALM BEACH GARDENS, FLORIDA

Tomorrow the Moon – The Story of the Space Age
by Joseph F. Alcock

Cover: "Buzz" Aldrin setting foot on the Moon, 20 July 1969 (courtesy NASA)

First published by Aerospace Publishers in 2016
Palm Beach Gardens, FL
Revised in 2025

ISBN: 978-0-9890111-2-9

Books by Joseph F. Alcock:

- *Tomorrow Has Wings – The Story of Captain Jack Alcock, First to Fly the Atlantic Non-Stop in 1919*
- *Tomorrow Is Supersonic – The Story of the Jet Age*
- *Tomorrow the Moon – The Story of the Space Age*

This book is dedicated to my lovely wife, Marilyn.

Contents

Preface

Acknowledgments

Prologue

Part One – Konstantin Tsiolkovsky

Chapter 1 - Beginnings

Chapter 2 - It's Possible

Chapter 3 - Recognition

Part Two – Robert Goddard

Chapter 4 - Dreams of Spaceflight

Chapter 5 - First Flights with Liquid Propellants

Chapter 6 - The First Space Race

Chapter 7 - Final Launches and Beyond

Part Three – Hermann Oberth

Chapter 8 - Another Visionary

Chapter 9 - The Girl in the Moon

Chapter 10 - The V-2

Chapter 11 - Man Into Space

Part Four – Sergei Korolev

Chapter 12 - The Gulag

Chapter 13 - The V-2 in Russia

Chapter 14 - The Road to Sputnik

Chapter 15 - Luna and Vostok

Chapter 16 - More Accomplishments

Contents (Continued)

Part Five – Wernher von Braun

Chapter 17 - The Telescope

Chapter 18 - On to Peenemünde

Chapter 19 - Operation Paperclip

Chapter 20 - Project Jupiter

Chapter 21 - Projects Mercury and Gemini

Chapter 22 - The Spirit of Apollo

Part Six

Epilogue

The Legacy

Bibliography

About the Author

End Notes

Preface

If you were asked, "Where did space technology begin?" what would be your answer? This is a story that clears up the general misconceptions of its beginnings, and how we got to where launch vehicles have put men on the Moon and frequently put satellites into orbit for communications, weather prediction, and military surveillance to name just a few applications.

The story is non-fiction for the most part. All the events contained herein are events that actually occurred. I have taken the liberty to write each section as an autobiography of five major players (there were others) to bring out the challenges they faced and to provide a more interesting story. Based on my research of the individuals, some of the dialog is a fictional account of what was probably said, although much of it is the actual conversation that occurred.

As in most stories of great men, the emphasis in the story is on the obstacles they encountered and how they overcame these difficulties. There is some personal information, but, like most historical stories, at least in the field of science and engineering, there isn't much detail on their personal lives. However, I did bring out their feelings of rejection, discouragement, and abandonment and the fact that they were even persecuted before being recognized for what they were – agents of the future.

In a story of this nature, I found it necessary to include some technical information and description, without getting too much into the technical details. For those who would like more detailed information, the notes at the end of each chapter are provided.

Joseph F. Alcock

Acknowledgments

Many thanks to the Abacoa Writers Group of Jupiter, Florida and to Lawrence G. Tanner, former chief engineer at Pratt & Whitney Rocketdyne. Also, thanks to my son-in-law, Christopher Kachouroff. All the encouragement and suggestions are greatly appreciated.

Prologue

"There are at present time, two great nations in the world, which seem to tend towards the same end although they started at different points: I allude to the Russians and the Americans. Both of them have grown up unnoticed…The American struggles against the natural obstacles that oppose him; the adversaries of the Russian are men: the former combats the wilderness and savage life; the latter, civilization with all its weapons and arts; the conquests of one are therefore gained by the ploughshare; those of the other, by the sword…Their starting point is different, and their courses are not the same; yet each of them seems to be marked out by the will of Heaven to sway the destinies of half of the globe."

- Alexis de Tocqueville, *Democracy in America*, 1835.[1]

Part One

Konstantin Tsiolkovsky

Chapter 1 - Beginnings

Earth is the cradle of humanity, but one cannot live in a cradle forever. - Konstantin Tsiolkovsky

Kaluga, 150 kilometers SW of Moscow, 1895

"KONSTANTIN TSIOLKOVSKY, have you lost your mind?" asked Vladimir, a fellow teacher at Kaluga. "This has to be the wildest idea I've ever heard of. Your great learning has affected your thinking to the extent I think you've gone mad. The very idea of a space elevator is preposterous."

"Vladimir, listen to me," I countered, as I sat at my desk in my classroom. "I looked at the Eiffel Tower and I imagined it attached to a celestial castle at the end of a spindle-shaped cable, with the castle orbiting the earth in a geosynchronous orbit. We could travel up and down the cable to go from Earth to space and return."

"There you go again, using words I don't understand. Tell me, what is geosynchronous orbit?" he said and shook the manuscript I had given him to read.

"It's where an object's orbital speed is the same as the Earth's rotational speed so that it stays in the same place on Earth. It's about 22,000 miles above the surface of the Earth."

"And what keeps gravity from pulling the whole thing down to Earth?"

"Centrifugal force. It's like tying a rope to a bucket of water, then swinging the bucket in a circle around your head. Just as centrifugal force keeps the water in the bucket, a large counterweight in space holds up the cable. Do you understand?"

"*Da, da,* yes, yes. *Kanyeshna,* of course. You certainly have a great imagination, Konstantin." Then, without saying another word, Vladimir just walked out of the room, shaking his head and leaving my latest scientific writing about space elevators on my desk.

<><><>

I WAS BORN ON 17 SEPTEMBER 1857 at Izhevskoye, on the Baltic Sea. It was part of the Russian Empire at that time. My father, Eduard Ignatyevich Tsiolkovsky, was Polish and my mother, Mariya Ivanovna Yumasheva, was Russian. We lived in several places, because Father was a forestry official, a teacher, and later a government official.

When I was nine, I caught scarlet fever and became hard of hearing. This prevented me from being admitted to elementary school. As a result, Mother schooled me at home. I didn't have many friends, being at home so much, so I developed an interest in books, which led to an interest in flight and model balloons. Mother and Father encouraged me to continue reading and experimenting with balloons.

Mother died when I was thirteen, and after a while, I began falling behind in my studies. One day, Father came to me in my room and said, "I'm concerned about your education, so I'm sending you to Moscow. I've made arrangements there so you can continue your studies."

At Moscow, I met Nikolai Fyodorov, who tutored me in the library almost every day for about three years. He introduced me to books on mathematics and science. I studied chemistry, mathematics, astronomy, and mechanics, and I attended lectures with the aid of an ear trumpet. From these lectures, I achieved a better understanding of the problems of flight.

I classified Mr. Fyodorov, a Christian philosopher, whom Tolstoy, Dostoyevsky, Leonid Pasternak,[1] and others admired.

As I sat on the lawn outside the library one day, Mr. Fyodorov sat down alongside me and said, "The human race needs to regulate nature and have a common task that could unite mankind."

"What do you think that task should be?" I asked, somewhat puzzled by his sudden conversation on so deep a subject.

"Natural phenomena will remain visions until they become the product of the general activity of all humans. And notions will remain ghostly and dream-like until they become projects to be achieved by the general population."

"What projects do think would be the most beneficial?" still perplexed by this encounter with Nikolai's philosophy.

"Death and after-death existence should become the subject of comprehensive scientific inquiry. Achieving immortality and revival is the greatest goal of science. We must be learning everything relevant to achieving that goal. Living in space may be the way to conquer death."

From Fyodorov's discussions about space, and the writings of Jules Verne, especially Verne's book *From the Earth to the Moon*,[2] I came to believe that colonizing space would lead to the perfection of the human race, with immortality and a carefree existence the ultimate result. Thus, space travel, to me, was simply a tool to achieve higher goals.

In 1876, Father came to visit me in Moscow. "I'm hearing that you are working too hard and are going hungry. Is that true?"

"Well, I'm totally absorbed in my study of mathematics and physics, and a student doesn't make much money," I answered.

"I think it best if you were to return home with me until you can find some work and not have to live like this," Father said and motioned around the small dingy apartment where I had lived for several years.

I returned to Kaluga and remained there until I passed the teacher's exam and, in 1878, found work as a math teacher at a school in Borovsk, near Moscow. It was there I first encountered a beautiful and intelligent young woman. I married Varvara Sokolova, the daughter of a local preacher. We had five children: three daughters and two sons. Varvara was wonderful, as she encouraged me to continue with my own scientific studies.

I carried out some experiments with steam engines, pumps, and fans in the laboratory, which I built in our apartment. "Its theoretical calculation that directs my thought and my imagination," I told Varvara.

In 1881, I sent the manuscript of my first scientific study, "Theory of Gases," to the Russian Physico-Chemical Society (RPCS) in St. Petersburg. In it, I outlined the basis of the kinetic theory of gases. Shortly after, the prominent chemist, Dmitry Mendeleyev,[3] informed me, "Shortly after, Dmitry Mendeleyev informed me, "We can't accept your work because someone submitted the same theory twenty-five years ago."

Initially, I felt very discouraged, but Mendeleyev inspired me to keep working on my research and submit another manuscript.

"The Mechanics of the Animal Organism" received a positive reception, and the Society inducted me. About this time, I dealt with four major scientific pursuits: an all-metal dirigible, streamlined airplanes and trains, hovercraft, and rockets for interplanetary travel. However, none of these received much recognition or support from the scientific world.

In 1883, I published *Free Space,* which described a space vehicle in some detail. It showed cosmonauts in weightlessness, gyroscopes for attitude control, and an airlock for exit into free space.

In 1889, workers completed the Eiffel Tower in Paris, making it the entrance arch to that year's World's Fair. It impressed me not only as the tallest man-made structure of the time, but it illustrated that the use of steel enabled the creation of a 334-meter tower. I wondered what else metals could accomplish. Hence, the idea of a space elevator entered my mind.

<><><>

VLADIMIR LEFT MY CLASSROOM that day in 1895 after we talked about a space elevator. He, like many of my colleagues, thought my ideas impossible. Certainly, the capability to put a cable into space isn't practical and the stresses in the cable would exceed what steel or any other material could withstand, but someday, yes someday, there might be a solution.[4] [5]

Thus, I turned my attention to solving some other problems, and in the next few years, I discovered something that would be the basis of all rocket designs.

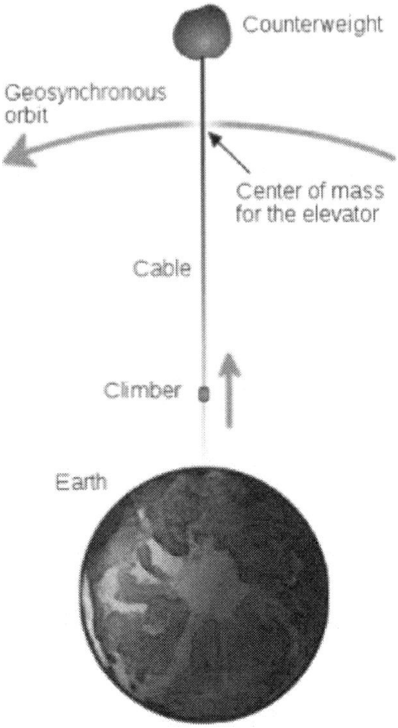

Concept of a space elevator
(Sketch in the Public Domain)

Chapter 2 - It's Possible

For a long time, I thought of the rocket as everybody else did–just as a means of diversion and of petty everyday uses. I do not remember exactly what prompted me to make calculations of its motions. Probably the first seeds of the idea were sown by that great fantastic author Jules Verne–he directed my thought along certain channels, then came a desire, and after that, the work of the mind. - Konstantin Eduardovich Tsiolkovsky

Kaluga, 150 kilometers SW of Moscow, 1895

DESPITE VLADIMIR'S CRITICISM, I wrote "Dreams of the Earth and Sky." I was intrigued by the idea of a mineral-rich Moon, one we could someday harvest. I believed it might contain gold usable for the benefit of humankind.

Again, after the students had left one afternoon, Vladimir came to my classroom. He had a copy in his hand of my latest work and proceeded once more to tell me I had an overactive imagination. "Konstantin, you keep going down this insane path of space travel, living in space, and mining the Moon. Wouldn't you be more productive if you did something else besides engage in these wild daydreams of yours? And another thing, how could you keep anyone alive in space?"

"We could construct a station in space, in the shape of a torus," I answered patiently. "We could make the torus rotate to create artificial gravity, so we would feel like we were back on Earth.

There could be plants in the station to give off oxygen to keep people alive."

"Whatever is a torus?" Vladimir responded impatiently.

"In this case, a giant rotating wheel where people could live along the rim. It could have collectors to capture the sun's energy and a device to send messages back to Earth. From this celestial castle circling the Earth, we could travel to the Moon someday."

"And how do you propose getting to and from your celestial castle?"

"Well, a device that expels a gas at a very high velocity could propel men and equipment into space. These are typically called reactive devices[1] and would be like a Chinese rocket."

"What would it look like, and how would it work? You haven't shown me anything that explains this," Vladimir said with raised eyebrows.

"I'm working on a paper to do just what you ask," I answered in a soft voice.

"Humph," was all Vladimir said as he walked out of my classroom.

<><><>

THE NEXT YEAR, 1896, I WROTE, "Exploration of Space by Means of Reactive Apparatus," and after that, I calculated how fast a rocket would have to travel to reach and stay in Earth orbit. A device would have to reach slightly less than 8,000 meters per second to achieve Earth orbit,[2] the exact speed depending on the shape and location of the orbit, and some other factors.

More importantly, on 10 May 1897, I derived what later was called the Rocket Equation, the basis of all rocket designs. It allows the designer to estimate the final velocity of the rocket when all the propellants are spent to determine if it could reach Earth orbit.[3]

While I was writing all this down at the desk in my workroom in our log home, Varvara called to me from the kitchen.

I was so excited about finding the Rocket Equation; I didn't hear what she was saying. Instead, I ran to the kitchen shouting, "It's possible! It's possible! It's possible!"

"What on earth are you talking about?" she asked.

"I now know what is necessary to reach Earth's orbit and stay there." Then, I explained the principles I discovered to Varvara, including the need for a very powerful fuel to reach orbit. "I can calculate a rocket's speed by knowing the speed of the gas at its exit, the mass of the rocket and the mass of its propellants."

"It sounds exciting," she said. "You need to publish it before someone else does the same thing.[4] I have a question, though. You talk about putting a device in orbit and it stays up there, circling the Earth. How is this possible since it doesn't have any fuel? Doesn't gravity pull it back to Earth?"

"Good question. I've asked myself that same question when I first began this work. An object, or satellite, that is thrown into the air falls back towards Earth in an arc because of gravity. A space vehicle also falls back to Earth in an arc when the reactive device runs out of fuel, but it's traveling so fast around the Earth that the centrifugal force acting on it pushes it out into space. When the centrifugal force equals the force of gravity, which acts in the opposite direction, the satellite stays in orbit around the Earth. Let me draw it for you."

<><><>

IN 1903, I SUBMITTED MY FINDINGS to the Russian journal *Nauchnoye Obozreniye*.[5] It appeared, together with many of my ideas on space flight, in an article called, "Investigating Space with Rocket Devices."

Shortly after the article appeared, my daughter came to me and said, “Father, *Nauchnoye Obozreniye* is being confiscated by the Tsar’s orders. The journal contains political statements against the Tsar. Wasn’t your space research published there?”

Unfortunately, the journal ran a revolutionary piece that led to its confiscation.

Vladimir, however, managed to get a copy of the journal, “Konstantin, your article is fantastic. I’m very impressed with it. Explain to me how a rocket could have enough energy to go into space.”

“Visualize an elongated metal chamber. The chamber is partly occupied by a large store of substances, which, on being mixed, immediately form an explosive mass. This mixture, exploding in a controlled and fairly uniform manner, flows in the form of hot gases through a tube with a flared end, shaped like a trumpet. The gases flow outward from the end of a tube into space at a tremendous velocity. Clearly, such a projectile will ascend like a rocket. We need substances that would produce a high exit velocity, such as a fuel with a high amount of energy from the explosion.”

“I understand what you’re saying, but what substances could explode with a velocity high enough to produce the thrust required to lift a rocket into space?” Vladimir questioned.

“A combination of liquid hydrogen and liquid oxygen might provide enough thrust, for the amount of propellants used, to allow a rocket to ascend into space,” I answered.

“That’s preposterous. How would you create liquid hydrogen and liquid oxygen?”

“At the present time, it is not difficult to convert hydrogen and oxygen to liquids. Their temperatures would be extremely low,” I answered.[6]

Other fuels probably could not produce the exit velocity from the rocket needed to reach space, so later on I worked on the rocket design itself to achieve the necessary flight velocity. I discovered, or should more correctly say, rediscovered a principle the Chinese used centuries before in fireworks design.

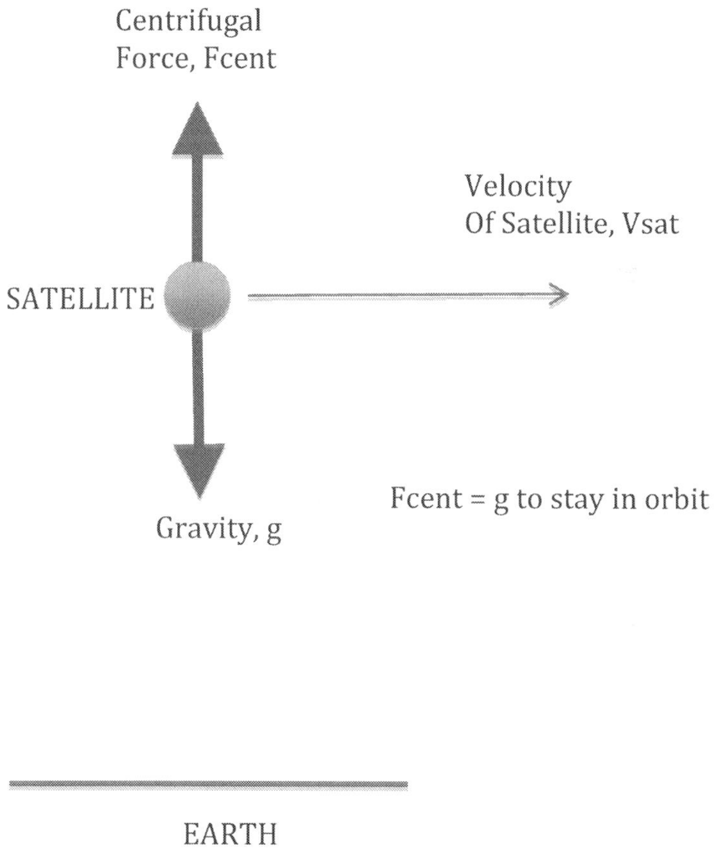

Drawing showing how a satellite stays in orbit around the Earth: To stay in orbit, the centrifugal force on a satellite must equal the force of gravity.

Chapter 3 - Recognition

Mankind will not remain on the Earth forever, but in the pursuit of light and space will at first timidly penetrate beyond limits of the atmosphere and then will conquer all the space around the Sun. – Konstantin Eduardovich Tsiolkovsky

Kaluga, 150 kilometers SW of Moscow, 1908

"*KAKOY KOSHMAR! KAKOY KOSHMAR!* What a nightmare! What a nightmare!" screamed Varvara as she came running into my workshop.

"What on earth is upsetting you so much?" I asked.

"The river is flooding! We need to leave immediately. No time to take anything with us."

Thus, I left everything behind as we ran for our lives ahead of the flood from the Oka River. Everyone in the area fled to higher ground.

When the flood subsided, Varvara and I returned to our home and found the flood had swept away most of our belongings, including much of my research work.

<><><>

THE TWENTIETH CENTURY had not been kind to us. Besides the flood, our son, Ignaty, committed suicide in 1902, and in 1911, my daughter got into serious trouble. One day, Varvara came to me at my desk and said, "I'm so weary, Konstantin. We've had nothing but disappointments and heartache. The harsh life we lead took Ignaty from us, and the tsar's men have arrested our daughter for her revolutionary activities."

I thought, I need to stay positive and provide hope for Varvara and the family, despite our problems and oppressive life. "Cheer up, Varvara. Better days are coming."

"How in the world can you say that?" Varvara said between sobs. "Just look at what they did to you. Back in 1892, when you got the promotion for a teaching position here in Kaluga, we thought things would be better than in Borovsk, but they haven't been. You did all that work on dirigibles and got nothing for your efforts. It came from your own pocketbook. We can't eat a wind tunnel or research papers."

"You must remember, Varvara, the Society gave me 470 rubles to build my wind tunnel and aerodynamics laboratory to test metal dirigibles."

"That's all they ever did. And if you remember rightly, my dear husband, they gave most of the money to that foreigner to develop his dirigible model, which you tested and found to be inferior to yours." Varvara sighed then added, "I'm sorry, Konstantin, I'm just tired."

I thought, I know you are right, Varvara. "I still think things will change. I know what I'm doing will be useful someday."

"Maybe someday the bureaucrats will see the good work you've done."

"I hope so." I doubt it, I thought. The tsar, the nobility and the bourgeoisie spare no expenses on court balls and theatrical performances but grudgingly give but a few rubles for scientific work. Most of the scientific work done here in Russia is with the scientists' own money.

"Yes, maybe someday," was all Varvara said.

I thought, in my case, not having a diploma and living so far from Moscow, my chances are even less of getting some financial support. I can't get copies of the works of other scientists here in Kaluga, even if I could afford them. But what bothers me the most is some of my colleagues don't understand and appreciate my work.

<><><>

THINGS DIDN'T GET BETTER, and I too became discouraged. However, I continued in my work and authored *Investigations of Outer Space by Reactive Vehicles* in 1911. Once again, I published my Rocket Equation, and this time it received a wider circulation in the scientific world. As before, I showed that gravity needed to be factored into the equation. In 1911, however, I estimated the effects of air resistance, and I studied flights that did not have a vertical ascent. I also defined the equation to determine the flight speeds to leave the Earth, Moon, or one of the planets.[1]

In 1914, I supplemented my previous work by writing five theorems of space flight. In many of my publications, I emphasized what is impossible today is possible tomorrow.

I didn't do much research during the Great War, as conditions were so harsh. However, toward the end of the war, I saw a glimmer of hope. The Revolutionaries were gaining strength. In 1917, they removed the tsar, and then the Bolsheviks took power in October of the same year. I wondered, what effect will all of this have on me?

<><><>

WITH THE BOLSHEVIK TAKEOVER came personal recognition. The Socialist Academy, later renamed the Communist Academy, elected me in 1919.

My old friend, Vladimir, visited me at my workshop in 1921.

"Konstantin, I'm glad to see you are finally getting some support."

"Yes, they actually gave me a pension so I could retire from teaching and continue my studies. There is so much to do, and now I can spend all my time on research."

"What are you planning to do?" he asked, as he looked over the projects I had been working on.

"Well, I've been thinking more about the problem of getting into Earth's orbit," I answered. "My previous calculations showed it possible, but they are under ideal conditions. I need to determine how to better design reactive vehicles."

"What are you so concerned about?"

"Weight. That is the big unknown. Will the design be too heavy?" I wondered out loud.

Vladimir, now a supporter of my work, simply said, "I know you will think of something."

<><><>

I WORKED ON MANY THINGS in the ensuing years and published many papers on various subjects, including all-metal dirigibles and jet aero planes, but the possibility of space travel was my passion. One particular article got the attention of many. It was part of a series of articles I published in 1929.

Vladimir once again visited my workplace. "I understand you solved the problem of getting to space. A rocket train seems a little, how would you say, extreme, though?"

"Yes, I know. By rocket train, I mean a combination of several identical reactive vehicles moving first along a roadway, then in the air, then in space beyond the atmosphere, and finally between the planets or suns. But only one part of this train reaches celestial space, the other parts, since they lack the necessary speed, return to the Earth."

"Whew," was all Vladimir said. After he thought some, he added, "You're talking about several rockets, one on top of another. That's complicated. Why not simplify the design and just have one rocket as you described before?"

"To reach orbital velocity, a single rocket would have to use a large supply of propellants. Thus, to attain orbital velocity, which is about 8,000 meters per second, the weight of the propellants has to be at least eighty percent the weight of the rocket with all its contents.[2] This amount of propellants complicates the construction of reactive vehicles. It would be difficult to design the vehicle structure, skin, tanks, and so forth within twenty percent of the total weight.[3] I've done some preliminary designs, which show the vehicle weight could easily exceed twenty percent. Now a train makes it possible to reduce the amount of propellants to relatively small percentages in each rocket, because each rocket doesn't need to provide a high final velocity. Each rocket imparts an initial velocity to the next."

I continued. "What you say, Konstantin, makes sense. I know the Party will act on it soon, as it is doing for ideas and discoveries of other Soviet scientists."

<><><>

BETWEEN 1925 AND 1932, I wrote about sixty works on astronautics, astronomy, mechanics, physics and philosophy. I spent considerable time studying rocket design and how people might survive in space. I am concerned about how high acceleration during a launch[4] and how weightlessness in celestial space affects people.

In September 1935, my health was not good. I was told I needed to go into the hospital for surgery on my stomach. It doesn't look like I'll ever build a rocket before I die.

Epilogue

KONSTANTIN E. TSIOLKOVSKY DIED after surgery for stomach cancer on September 19, 1935, at the age of 78. The Soviet government gave him a state funeral, and they buried him in the old Kaluga Cemetery near the museum honoring his life and work.

Tsiolkovsky never built a working rocket himself, but he influenced many Russian engineers and designers who made his visionary concepts a reality. Among them was Sergei Korolev, who became chief designer of the Soviet Space Program and the driving force behind the Soviet space successes in the 1950s.

"Time on occasion inexorably erases the outlines of the past, but the ideas and works of Tsiolkovsky will more and more attract attention as rocket technology proceeds in its development," said Sergei P. Korolev in 1957 on the centennial of Tsiolkovsky's birth. "Konstantin Eduardovich Tsiolkovsky was a man who lived far in advance of his times as befits a true and great scientist." [5]

A far side Moon crater is named in his honor. In 1989, he was invested in the International Aerospace Hall of Fame.

Three men are generally credited with being the most prominent pioneers of rocketry and spaceflight theory. One was an American, Robert H. Goddard, and one was a German, Hermann Oberth. Both of these men made substantial contributions to science, but they came on the scene after the pioneer Konstantin Tsiolkovsky had already begun seriously thinking about leaving "the cradle of the mind" – Earth!

Konstantin Eduardovich Tsiolkovsky
(1857-1935)
(Photo from the Public Domain in Russia)

Part Two

Robert Goddard

Chapter 4 - Dreams of Spaceflight

It is difficult to say what is impossible, for the dream of yesterday is the hope of today and the reality of tomorrow. – Robert H. Goddard

Clark University, Massachusetts, 1919

"I DON'T WANT TO DO IT!" I told Dr. Webster. "I believe high altitude is possible and practical. I want to protect my research from copycats and rivals."

"Dr. Goddard, what good is all your research if no one knows about it? Besides, there are ways of protecting your inventions, which you know about."

Arthur Webster was not only a brilliant mathematician but a good friend as well. However, on the issue of publishing, we had major differences. Arthur set down all his experiments and theories in various writings. Here in his workshop, he had blackboards filled with his notes, piles of lecture outlines on his desk and articles he had published in scientific journals stacked in bookshelves with other articles and books he had collected. He wasn't the least bit reluctant to share his work with anyone.

"I still don't want to give my work away to someone or some company who hasn't spent all the time I have formulating methods to reach space," I said in protest.

"As head of the physics department, and your boss, I might add, you have a choice: either you publish your work, or I'll do it for you. Of course, I'll give you the credit for what you've done," Dr. Webster said. He was quite a commanding figure, and right in style, with his thick, dark moustache. Few disagreed with him.

Having no choice, I convinced him to allow me to publish a modified version of my article "A Method of Reaching Extreme Altitudes," which I had written several years earlier.

Dr. Webster sent the paper to the Smithsonian Institute, and one particular paragraph, which I had added, created an uproar there and elsewhere.

<><><>

I WAS BORN IN 1882 IN WORCESTER, MASSACHUSETTS, to Nahum Danford Goddard and Fannie Louise Hoyt. I was their only child, as my younger brother died before his first birthday. As a boy, I was curious about nature. I studied the heavens using Father's telescope, and I observed how birds fly. I loved the outdoors and became an excellent marksman with a rifle.

I was often sick, usually with bronchitis, which kept me home from school. To pass the time, I read my father's copies of *Scientific American*, and the new inventions described in that magazine fascinated me. There were articles on the latest radios, automobiles, and phonographs.

My mother also had health problems. She caught tuberculosis, and the doctor advised our family to move back to Maple Hill in Worcester, about 40 miles west of Boston, so she could recover in the cleaner, drier climate than at Roxbury by the sea.

At Maple Hill, I once again caught bronchitis and spent my time reading library books, which I soon tired of. However, Father had the *Boston Post* delivered every day, and I became engrossed in a story by H.G. Wells, "The War of the Worlds." A new chapter appeared in the *Post* every week. It was about an army of Martians who came to Earth in a spaceship to colonize Earth for their purposes. It was a fantastic story, not believable, but it made me wonder, how did that spaceship work? I also read the works of Jules Verne, especially *From the Earth to the Moon*. I read these stories many times because they fascinated me so much.

I thought about what it would take to escape the Earth's gravity, a rocket that could go 25,000 miles per hour. The only possibility, I thought back then, was to use plenty of gunpowder to go that fast.

I was a good student and especially good at math and physics. I quickly came up with the answers to the problems assigned to us. As a result, Mr. Andrews, my high school physics teacher, took a liking to me, and we became good friends.

One day after school, we met in the library. I had many questions about spaceflight to ask Mr. Andrews. I remember asking him, "How could one design a rocket to get into space?"

"I don't know. In fact, I don't think anyone can answer your question–at least not right now. Someday, there may be an answer."

I thought the reaction of burning fuel escaping from a device throughout the flight was probably the only way to reach space. Gunpowder was the propellant used in most rockets, but it exploded all at once, which made it unsuitable for a rocket intended to reach space. So then, what fuel would burn slow enough in order to accelerate a vehicle continually for it to reach escape velocity?

It would take some time to answer that question. When I graduated from South High School in 1904, I still found no answers. Therefore, I felt very discouraged about spaceflight. None of the books I studied provided any answers. Most of them stated spaceflight was impossible. I reacted to this by tossing all my notes on the subject into the oven and forgetting about space for a while.

My father couldn't afford expensive schooling, but my grandmother, Gram, as I called her, borrowed money so I could attend Worcester Polytechnic Institute. I was a general science major. There was no mention of spaceflight in any of my classes, however, an understanding of physics and mathematics would be very useful later.

But I couldn't forget my dream of spaceflight. I worked on using power from the Sun and modifying an artillery rocket. Gunpowder rockets had been around for centuries, but they burned uncontrollably and didn't have the power or accuracy of big guns. The need for the right fuel to reach space became apparent. My chemistry books seemed to indicate liquid hydrogen would burn steadily and liquid oxygen could take the place of air needed for combustion to occur. These were dangerous and expensive substances and would be difficult to use in any experiment.

During my undergraduate days, I wrote an article on gyroscopes, which impressed my teacher to the extent he recommended I submit it to *Scientific American* for publication, which they did. However, articles I submitted on spaceflight were rejected on the grounds they were fantasy.

<><><>

AFTER GRADUATION, I ENROLLED at Clark University in Worcester and earned a doctorate in physics three years later, in 1911. The university appointed me an honorary fellow, granting me laboratory access but no salary. I researched methods for measuring electric current in vacuum tubes and developed a measurement device, which I called a vacuum tube oscillator.

When the officials at Harvard heard about my research, they invited me to give a speech about my work. After my talk, Dr. W.F. Magie, the dean of Princeton, came up to me. "Young man, would you be interested in a job at Princeton?"

I was glad to accept the job as it gave me ample time to do research, and, most importantly, I received a salary. I began work there in the fall of 1912 and continued my electrical experiments. Once again, I was free to work on the calculations needed to build my first rocket.

Unfortunately, my new freedom vanished in the spring of 1913, when I visited my parents in Worcester. I began coughing and had a high fever. Father had the family physician, Doctor Fischer, look at me, who called in Dr. Lapham a tuberculosis expert. He said, "Robert, you have tuberculosis, and it's very serious. It has spread throughout your lungs. You need to move to a drier climate and stay in bed if you are to recover."

I didn't know it then, but he gave me about two weeks to live. Instead of following his advice, I remained in Worcester and built up my strength by taking walks. Each day, I walked farther to increase my strength. After about a year I recovered.

Dr. Lapham told me, "If you go five years without the disease reoccurring, you'll probably never have it again. You'll be better than you ever were."

I avoided discussing my love of cigars with the good doctor, and I returned to Clark University to teach electrical courses. As new as this science was, I ventured off into the newer science of rocketry.

I ordered a variety of rockets delivered to my lab. There were Chinese rockets, British Congreve rockets,[1] signal rockets, fireworks, and flares. They all used gunpowder for their propellants. I examined how their shape affected their flight.

But these rockets never got very far off the ground. They were heavy, hard to control, and they burned inefficiently.[2] I designed and tested several. One reached an altitude of 486 feet.

Not wanting my ideas stolen by another inventor or by a commercial business, I went to Charles Hawley, a patent attorney in Worcester. After seeing my sketch, he said, "Robert, your rocket design has merit."

Charles drew up the patent application and sent it in. Before the patent office reviewed it, I had other thoughts on rocket design. I went to Mr. Hawley's office in Worcester several times to apply for another patent. "Your rockets are quite different from your previous patent application on electricity," he said. Tell me, in simple terms, what you think is significant about these designs."

"There are three broad principles in these patents: the use of a nozzle attached to a combustion chamber for additional thrust; the feeding of successive portions of propellant, liquid or solid, into the combustion chamber; and the use of multiple rockets, each discarded in succession as the propellant it contains is exhausted."

Thus, Mr. Hawley submitted two patent applications under the same title, "Rocket Apparatus," and the patent office awarded me the patents in 1914.[3] I perfected these designs and filed for additional patents.

<><><>

I DID ALL THIS WORK using my salary to fund it. I needed more funding to continue, so, in 1916, I went to the Smithsonian Institute in Washington, D.C. for help. I wrote an article entitled, "A Method for Reaching Extreme Altitudes," which contained my calculations, showing it was possible to send weather recording devices to altitudes ten times higher than research balloons could reach. Dr. Abbot, of that institution, was very interested in high altitude research and recommended to Dr. Walcott that they fund my work.

The following year, the Smithsonian gave me a grant for $5,000 and the Army Signal Corps gave me one for $20,000. The money went to a small workshop at Worcester Poly. About the same time, the U.S. entered the Great War, and my superiors directed me to develop rockets as weapons. I and the military were worried about enemy spies, causing me to cover the windows of my workshop with blankets and hire a night watchman to maintain secrecy. Still concerned, with the help of the military, I moved my equipment to Mount Wilson, near Los Angeles.

By November 1918, my weapons were ready. I had developed a

rocket launching tube soldiers could easily handle. Clarence Hickman, my assistant, and I demonstrated it at an army firing range. It was successful, completely destroying the targets and imparting little recoil.[4] The generals and admirals approved of the results and promised continued funding for my research. A few days later, Germany surrendered, and the money stopped.

<><><>

THE SMITHSONIAN RECEIVED MY REVISED ARTICLE on reaching extreme altitudes and published it in December 1919. One paragraph, which I had added to the original, was a thought I had about the future of spaceflight. I wrote that someday a rocket carrying a gunpowder charge could be sent to the Moon, detonated on arrival, and its explosion visible from Earth. Scientists were the primary audience for this article, as only they could understand the physics and advanced mathematics. However, the Smithsonian sent a notice to several newspapers, which created considerable excitement.

The notice was very sensational, to say the least. Every major newspaper received it. *The New York Times* wrote: AIM TO REACH MOON WITH NEW ROCKET. I was dubbed the "Moon Rocket Man." Reporters called inquiring about spaceflights, and people volunteered to be on the first flight to the Moon. Others were not so supportive. Some scorned my work, claiming a flight to the Moon was impossible.

I was annoyed that people were so ignorant, but I saw it as an opportunity to raise some money for my research. Not much came from my fund-raising efforts as the publicity died out and people forgot about my work. I had spent the grant money from the Smithsonian; therefore, I was relieved when Clark University gave me $3,500 for two more years of research.

The disclosure of my work by the Smithsonian precipitated an unusual proposal. Hermann Oberth of Germany had read my article and wrote a letter to me in May 1922, offering to work with me to develop a rocket. I politely refused. However, he continued on his own, and the next year, he published *The Rocket into Planetary Space*.

Dr. Webster stopped by my office just after I had read Oberth's treatise. "What do you think of his work?" he asked.

"That blasted Hun[5] stole many of my ideas! On top of that, the acknowledgement he gave me was in the appendix, which said I was a mere theorist and didn't see the possibilities of spaceflight. I told you we shouldn't have published my work."

"It's apparent that we're in an international space race."

"Yes, and I need to reevaluate my work completely."

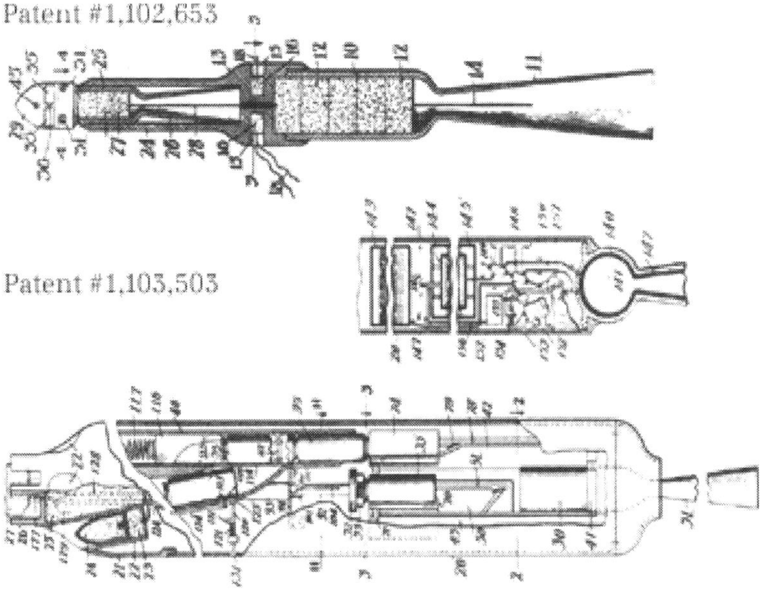

Drawing from Robert Goddard's 1914 patents for a two-stage solid propellant rocket (top) and a liquid propellant rocket (bottom). (From NASA)

Chapter 5 - First Flights with Liquid Propellants

There can be no thought of finishing, for aiming at the stars, both literally and figuratively, is the work of generations, but no matter how much progress one makes there is always the thrill of just beginning. - Robert H. Goddard

Clark University, Massachusetts, 1924

THINGS IN LIFE ARE NEVER CONSTANT. And so it was with me. I lost Gram in 1916 and Mother in 1920. Then Dr. Webster forced me to publish my work, which caused me more grief over all the negative publicity. Now, Oberth was stealing my rocket designs. However, things were changing for me.

I had met Esther in October 1919. She was a typist at the president's office at Clark. She had graduated South High School the previous spring and was working to earn money for college. I had many notes to be typed, so I saw her quite frequently. She was very intelligent and physically very attractive. Her parents were from Sweden, and she certainly had a Scandinavian appearance. She was a blue-eyed blond, slender with a high brow, straight nose and firm chin. She reminded me of Mother. I was absolutely smitten by Esther and saw her as often as possible.

I took her riding in my jalopy, an Oakland roadster, which is all I had left from a consulting firm Thompson, Riffolt, Hickman and I had founded. The company failed because of lack of business.

One day Esther said, "Robert, you need to take a break. You're in that lab constantly. Let's spend some time together on weekends."

We took walks on Sunday afternoons and stopped for ice cream. We developed an interest in photography, and she encouraged me to take up painting. I loved nature and painted many landscape scenes. Often, I would play the piano for her.

"You play the piano by ear. Not bad for a physicist," she once said. "You are very precise in your music. Are you really playing by ear, or are you following an equation for that, too?"

Those were such wonderful days. We were so naïve about life and very much in love. Unfortunately, her parents insisted she continue her college education. At Bates College, in New Hampshire, she earned straight As in English, French, German, Greek, history and public speaking. I missed her, but she was very intellectual and needed to continue college. She was interested in cultural things, books and theories. She read a lot and was an interesting talker, as my friends described her.

When we became engaged in 1922, she announced, "I'm old-fashioned. I'm dropping out of college to get things in order for our marriage." She returned to her typing job to save up for her trousseau. By June 1924, she was ready, and on the 21st of that month, we had a very simple wedding at St. John's Episcopal Church in Worcester. It was the beginning of a wonderful relationship.

<><><>

BECAUSE OF OBERTH, I was determined to build and launch a rocket as soon as possible. Realizing solid fuels would never reach high altitudes, because of their weight, I concentrated on liquid propellants. For ten dollars, I could buy two liters of liquid oxygen from a local welding shop, and because liquid hydrogen was unavailable, I used gasoline as a fuel instead.

I strengthened my design by using duralumin, a metal made from copper, magnesium, manganese and aluminum, for the combustion chamber. It was lightweight and could withstand the temperatures that would cause other metals to melt.

"Henry, we're ready," I told my assistant, Henry Sachs, in March 1926.

"It's been a lot of work. We've checked everything for leaks and cracks, so when the weather improves, we should be ready to launch," he said.

I had made arrangements with a distant relative, Effie Ward, to use a large empty field on her farm to launch my rocket. Her farm was at Auburn about five miles from Worcester, so the noise from my testing wouldn't be a problem.

Henry and I hauled the 10 ½-foot rocket out to the farm. It wasn't very heavy. Even loaded with propellants, it weighed only about ten pounds. It was cumbersome, as it was completely different than my later designs. It had a frame of metal tubes, which connected the combustion chamber and nozzle at the top to the propellant tanks at the bottom. Because it lacked an outer case, all the parts were easily visible.

Esther was very supportive of it all. "I'll join you at the farm and bring a movie camera to record the tests," she said.

We made an attempt on March 6th, but the nozzle burned off, and the rocket jammed into the frame and bent it.

After some design changes, we tried again on March 16th. Before the attempt, Esther said, "Stand alongside it. I'll take a picture of you."

So, I posed by the rocket in its frame and Esther took a picture. Then I said to Henry, "Go ahead."

Henry poured the propellants into the tanks, and Esther stepped back to take a movie to record the event. Henry lit a blowtorch attached to a long stick and touched off the igniter at the top of the motor. He got behind a wooden door for protection as I opened a valve that let the oxygen and gasoline flow into the combustion chamber.[1]

Although the valve opened, the rocket didn't rise at first, but the flame came out and there was a steady roar. After several seconds it rose, slowly, until it cleared the frame, and then at express-train speed, curved over to the left, and struck the ice and snow, still going at a rapid rate. It looked almost magical as it rose, with no appreciably greater noise or flame. I didn't know what to expect, so I yelled, "I think I'll get the hell out of here," as I ran for some shelter.

The rocket weighed about 10 pounds but only created about nine pounds of thrust. Thus, the rocket didn't rise for 20 seconds, until the excess propellants burned off. It rose 41 feet and landed 184 feet away. The entire flight lasted only 2 ½ seconds.

Esther came over afterward and said, "The camera ran down before the rocket rose. I went to use the still camera, but the rocket had landed before I could get a picture of it in flight."

I contacted Dr. Abbot at the Smithsonian, who had funded some of this work, "The average speed from the time of the flight, as measured by a stopwatch, was 60 miles per hour. This flight was significant, as it was the first time a rocket operated by liquid propellants traveled under its own power. I wrote: "I'm requesting the Smithsonian keep this matter quiet until I complete my next step. My reason is that this rocket work is being made almost a national issue in Germany."

Dr. Abbot didn't say much in his response, as he probably wasn't very impressed. By comparison, balloons, which Dr. Abbot was very familiar with, were reaching altitudes of 25 miles.

On April 3rd, I launched again. It flew for 4.2 seconds before landing about 50 feet away from the launch tower. The nozzle was half burned through near the top, but the chamber was OK. On the 22nd, I tried again, but there was no lift. The chamber burned out near the bottom after about one minute.

Upon examining the rocket, I determined if I moved the combustion chamber down below the propellant tanks it would steady the rocket and perhaps lengthen its flight. On the 4th and 5th of May, I tried again with the new configuration but could not get lift. I now knew to reach high altitudes. I needed to increase exhaust velocity. I explained this to Dr. Abbot and told him, "I need to feed the propellants to the combustion chamber under pressure and to carry more propellants. This will make the whole thing bigger." I was now out of money, so I asked Dr. Abbot once again for more funding.

Dr. Abbot agreed the rocket must be larger, but he and Dr. Wolcott were skeptical. Abbot asked, "hat funds are needed to make a sufficiently large rocket immediately, so it could surely fly to a high altitude for a conclusive, spectacular demonstration. The Smithsonian can provide financial support if there is a commercial application."

"A high-altitude rocket would cost $2,500, but I can't see an immediate commercial use," I answered.

Dr. Abbot continued to press me, "We should have a really high flight to clinch the matter and not a crash. Before I authorize you to continue, I want to know exactly how you would bring about a spectacular flight."

We went back and forth before Dr. Abbot began funding me again. Sachs and I designed a new combustion chamber eight inches in diameter and over ten feet long. We worked on new valves, pumps, flow regulators and tanks. We went to what I called multiple feeding of fuel and liquid oxygen, LOX, by manifolds that distributed the propellants more evenly across the top of the combustion chamber. This proved most useful in future designs.[2]

I acquired a large windmill tower on Aunt Effie's farm to secure the rocket prior to liftoff. Dr. Abbot visited me at Thanksgiving in 1926 and was impressed with my lab and launch site. He asked, "Can you report a successful launch before President Coolidge attends a special meeting at the Smithsonian on February 11th?"

I told him, "Don't count on it. There's so much to do between now and February."

We moved the rocket to the tower in April. We tried to launch on April 16th and 20th, but igniter and fuel pressure problems prevented it.

On May 3, 1927, we again attempted to launch the 150-pound rocket. It lifted a short distance from the ground, but as it cleared the tower, the gasoline tank exploded.

I conducted several more tests through the summer, but none were successful.

"I still believe you can achieve high altitude with a rocket," wrote Dr. Abbot. "You just need to find the solution."

I appreciated Dr. Abbot's confidence in me. From his work on other projects, I think he knew, as I did, that sometimes we learn more from failures than from successes. I thought the best way to improve the reliability was to reduce the number of parts, as any good engineer would do. The new design was 12 feet long and weighed 32 pounds.

During the fall of 1928, I made additional launch attempts, and each time the combustion chamber burned out or some other failure occurred with the rocket never leaving the ground. I devised a method to cool the chamber. Injecting excess gasoline along the walls, with little oxygen to minimize combustion, seemed to work.[3]

On July 17, 1929, the rocket was again ready for firing. It cleared the tower this time and went to an altitude of 100 feet before the gasoline tank exploded. Neighbors heard the noise several miles away, leading them to believe it was a plane crash. They called the police to investigate.

As Esther and I examined the remains of the rocket, the police, accompanied by several reporters, arrived on the scene. “What’s going on?” they asked. They looked at the wreckage. “What’s this?”

I explained it was a rocket that had reached 100 feet before exploding. The next day, the newspapers said I had launched a rocket intended to reach the Moon. As a result, we had hundreds of people come to the farm to see another launch. Once again, the newspapers were writing about “flights to the Moon.”

The people of Auburn, worried about my rockets landing on them, asked me to find another place to do my testing. I packed up my equipment and moved to Camp Devens, an artillery range near Worcester.

I hoped I could maintain my privacy at the new location, but it was not to be. With all the publicity, there was a very positive outcome. I would get considerable support from sources I never expected.

Robert Goddard next to the first liquid propellant rocket to fly. (From Great Images In NASA, GPN-2002-000132)

Chapter 6 - The First Space Race

Just remember, when you think all is lost, the future remains. - Robert H. Goddard

Massachusetts, 1929

I RECEIVED A LOT OF PUBLICITY from the explosion of my rocket over Aunt Effie's farm. Some of it was over-optimistic about the future of my work, and some of it was quite negative and critical. However, there was some publicity I thought quite amusing.

In 1929, science fiction writer Philip Francis Nowlan, with cartoonist Dick Caulkins, created the comic strip *Buck Rogers in the 25th Century*, a story about a future space explorer. The optimistic scientific genius and friend of Buck in the story, Dr. Huer, looked very much like me. He invented Buck's rocket ship and other wonders in the story. Esther picked up on this. Whenever one of my experiments went wrong, she would quote Huer, "Now don't worry, the old doctor will take care of that."

<><><>

I WAS WORRIED as Dr. Abbot was providing minimal funding. I needed more money to purchase parts and pay my assistants. However, out of the blue came Charles Lindbergh to my rescue. Colonel Lindbergh had read about my testing at Aunt Effie's farm and was interested in "flying beyond the Earth," although he was told such flights were impossible. He decided to find out for himself and called me for a meeting.

In November 1929, he visited me at my lab, and I showed him my liquid propellant rockets, including the ones that had flown. I also showed him photos and movies of my experiments.

Lindbergh was impressed and asked, “Is it possible to send a rocket to the Moon?”

“Yes, it is possible with a multi-stage rocket, but it would cost at least a million dollars. That kind of money is out of the question.”

“How much do you need to continue your research, and what would you accomplish?”

“I would like $25,000 per year for four years. With that amount I could perfect high-altitude rocketry.”

Colonel Lindbergh, Slim as his friends called him, was a keen-minded young man and impressed me favorably. He was famous for being the first solo flight across the Atlantic in 1927, which gave him a lot of influence in many places. He was very supportive of my work and arranged for a meeting with the DuPont Chemical Company to get their technical and financial support.

The engineers and chemists questioned me in great detail to the extent I thought they were trying to steal my designs. They said, “Liquid-fuel rockets are impractical because the combustion chamber requires a heavy ceramic lining to withstand the heat.”

“DuPont is hopeless. Let’s go home,” Slim said.

Colonel Lindbergh flew me to Worcester. Later on, Esther asked me, “Doctor G, how was the flight home?” She usually called me “Doctor” or “Doctor G.”

"It was my first trip in an airplane, and I'm going to make it my last," I answered.

Through the winter and into the spring of 1930, Colonel Lindbergh worked with various groups to obtain funding for my research, but could only obtain a small amount of money. Finally, in May 1930, Lindy called me about a pledge from Daniel Guggenheim. I was so excited I took Esther out for dinner at a Chinese restaurant to celebrate.

"Lindy has had several meetings with Daniel Guggenheim, who pledged $100,000 to my work out of his own pocket," I told Esther at the restaurant. "He's guaranteeing $25,000 per year for the first two years. We can continue our work."

"What do they require of you?" Esther asked.

"After two years, I have to report my results to a committee of nine men, including Lindbergh, Dr. Abbot of the Smithsonian and Dr. Atwood, our controversial president of Clark University.[1] The Guggenheims also will follow my progress."

<><><>

WHILE LINDY WAS FINDING FUNDING for me, I continued my work. By early December 1929, I conducted several tests with two failures and one success. The success had a bulged chamber but was undamaged inside. By Christmas, I tested a wider chamber with gasoline spray along the inside wall, which I called curtain cooling. This rocket achieved 150 pounds of thrust before it exploded. I had mixed results in the summer of 1930. All this took place while I was working with Lindy and others to obtain more funding.

During this period, we got some bad press, and Larry Mansur, one of my assistants, came to me saying, "They ain't done right by our Nell." He was referring to Edward B. Sheldon's *Salvation Nell*, a Broadway play about a downtrodden girl of that name. The name stuck. During all our subsequent tests, Esther and I always referred to the rocket as Nell. To us, Nell became part of our family, as we had no children.

In July 1930, the district fire inspector visited my launch site and later reported my tests were safe, but the press negatively reported on the burned area at the base of the launch tower. I went to the state fire marshal to fend off these negative comments. However, on August 1st he forbade further flight tests in the state of Massachusetts. For this reason, combined with the poor roads in the area, the wretched weather, the mosquitos, and the unwanted publicity, we moved to a better location.

As long as we were moving, I decided to find an ideal site. After consulting with Lindy and a meteorologist, I settled on New Mexico because of its dry warm climate, large open spaces, and it was away from the public eye. Scouting the area, we finally came upon a deserted adobe house called Mescalero Ranch, near Roswell. We rented it from "Miss Effie" Olds, who no longer needed it after her children became adults and moved on.

My entire crew came with me to New Mexico. Mr. and Mrs. Henry Sachs, Mr. and Mrs. Al Kisk, and the bachelor Charles Mansur lived at the ranch, while Charles' brother, Larry Mansur and his wife lived in town, where she worked. We built a static test stand and workshop near the house, and Oscar White, a local rancher, let us use nearby Eden Valley on his ranch for our launch site, where we built a launch tower and control shelter.

I told Esther, "This site is far better than any in Massachusetts, but we aren't without problems. The greatest is finding a dependable supply of liquid oxygen, or LOX, as the men preferred to call it."

The Linde Air Products Company in Amarillo, Texas did what it could, but it wasn't enough. Dr. Abbot tried to locate a plant for me, but he was unsuccessful. Many plants only operated part time because of the Depression. I finally used nitrogen, instead of LOX, to pressurize the propellant tanks, which reduced the amount of LOX required.

In late October 1930, we conducted a series of static tests with some successes and some failures. By late December, our rocket was ready. It was eleven feet, including a nose cone, which contained a parachute that would release when the rocket stopped climbing. Below the cone, there were three tanks, the top one contained gasoline, the bottom one contained LOX, and the one in the middle contained pressurized nitrogen.

The pressurized nitrogen forced the propellants into the chamber at a high pressure for increased exhaust velocity. The nitrogen eliminated less reliable pumps for high chamber pressure. Four lightweight duralumin fins or vanes provided some flight stability. The entire system was open so we could see how the components performed after ignition.

On December 30th, Nell's launch occurred. My crew asked, "How did she do?"

"It went about 2,000 feet up and 1,000 feet away; time of ascent for 7 seconds," I answered. "I estimate it reached about 500 miles per hour."

"This is a first," said Esther. "None of our rockets ever reached this altitude or speed."

Later on, after I had some time to review the launch, we discussed the flight in more detail. "The engine put out a lot of thrust,[2] which caused the rocket to wobble in flight." I told the crew. "Also, the combustion chamber sustained some damage. These are areas we need to work on. The parachute's partial opening caused little damage, and we can easily fix that problem."

By September 1931, my rockets had a smooth casing to reduce the aerodynamic drag. I launched them using an electrical system to operate the valves before liftoff instead of using ropes. The flights weren't very successful, but the launch system and the combustion system were.

At that time, I began experimenting with gyroscopic guidance and made a flight test of such a system in April 1932. A gyroscope mounted on gimbals electrically controlled steering vanes in the exhaust. The rocket crashed after a short ascent. The guidance system had worked, based on the temperature of the vanes. I considered the test a success, although the vanes were too small for adequate control.

My work bothered Charles Mansur with gyroscopes. Knowing Lindbergh wanted a rocket to go high into the air in order to gain more financial support, Mansur and the crew said several times, "Let's get one way up in the air."

I answered them each time with, "Let's get it to go straight first. I'm not giving up on gyroscopes."

Mansur passed his thoughts on to me, "If you increase the exhaust velocity, which will increase the rocket's acceleration, the rocket will become self-stabilizing. Theodore von Karman [3] at the Jet Propulsion Laboratory and I agree on this." [4]

I didn't agree with Mansur and von Karman, and I continued to develop a gyroscopic system for steering.

But I had more than technical issues to deal with. The Great Depression caused a temporary loss of Guggenheim funding, forcing me back to Clark University from the spring of 1932 until the fall of 1934, when funding resumed.

While at Clark, I was worried that someone might steal my inventions, so I visited my old friend and lawyer, Charles Hawley.

"So, Bob, what brings you here today?"

"I know my patents of 1914 have expired, but I have several other rocket designs I want to protect. First, I've devised a method to cool the inside of the chamber, which I call curtain cooling, and second, I have developed a unique method of steering the entire rocket."

"Okay, is there anything else?"

"As a matter of fact, there is. I think some rocket principles can apply to aircraft. I have what I call a resonant chamber that produces a high exhaust velocity by using air that has been compressed and fed into a combustion chamber which exits at a high velocity." [5]

I showed Charles drawings of these devices with some detailed explanation, and he thought we should apply for patents. We did, and the patent office approved them. The downside was that they appeared in print in the United States and Germany. I knew Germany and the Soviet Union had rocket clubs and government projects trying to beat me into space.

<><><>

UPON MY RETURN TO ROSWELL, I began work on my A-series of rockets, 13 to 15 feet long, and powered by gasoline and liquid oxygen pressurized with nitrogen. I housed the gyroscopic control system in the middle of the rocket, between the propellant tanks.

In January 1935, the A-1 and A-2 failed to fly and the A-3 had limited success as the chamber burned through.

I had some successes and some failures, but A-8 rose 7,500 feet in May and A-10 6,600 feet in July, the highest ever up to then. After these flights, I did what I always did after a success. I lit up a cigar immediately, instead of waiting to have a smoke after I had examined the wreckage.

As I finished the A-series, the Smithsonian demanded, “You need to provide a model of your latest work and a full report” was the directive from Dr. Abbot.

We assembled a model from the wreckages and sent it to the Smithsonian. They agreed to keep it secret by putting it behind a screen away from the public. In March 1936, I published, “Liquid Propellant Rocket Development,” a ten-page report covering my work from 1930 to 1935.

I was sent a proof of the paper, but I never received it. Someone also opened several of my letters and packages before they arrived at my Roswell mailbox. I told Esther, “I suspect foreign agents are responsible for this monkey business.”

In 1936, the U.S. military attaché in Berlin asked Lindy to visit Germany and learn what he could of their progress in aviation. On his return, he met me at Roswell, “The Luftwaffe showed me their factories and were open concerning their growing airpower, but they were silent about rocketry.”

“Yes, they must have plans for the rocket,” I said. “When will our own people in Washington listen to reason?”

Robert Goddard at his Roswell, New Mexico workshop, 1935. (From National Air and Space Museum, Smithsonian Institution)

Chapter 7 - Final Launches and Beyond

I feel we are going to enter an era comparable in its progress to that in which the airplane advanced. It's just a matter of imagination how far we go with rockets and jet planes...I think it's fair to say you haven't seen anything yet. - Robert H. Goddard

Worcester, Massachusetts 1980

I'LL NEVER FORGET THE FIRST TIME I met Robert Goddard. It was at Clark University in October 1919. I was earning money for college by working as a typist in the president's office, having graduated high school the previous spring. He appeared one day and asked, "Esther, would you type some of my notes?" I did. He came often, and I liked him more and more as time went on. He was a tall, slender man, but had a very gentle and kind way with him. Soon we were seeing each other after hours.

We took rides in the country in his jalopy, as he called his automobile. It was a wonderful time in my life. He came to my house frequently with his notes to be typed. He used a paper punch to organize his notes, and my mother would get upset if we left any paper on the floor. She didn't approve of us as a couple, because he was so much older than me. He was almost forty, and I was only eighteen.

Nevertheless, we continued to see each other, and I asked him to take a break from his work on weekends so he could be with me. He loved nature, and on Sunday afternoons, he would paint landscapes, a colorful tree, a cloud or a milkweed pod. Inside, he would play the piano for hours.

As I got to know more about his work, I was in total disagreement with most of those in his profession. They often called his work "an impossible flight of fancy." I thought his rockets had possibilities, but I didn't know what it would take to make them work the way he wanted—to reach high altitudes.

I went away to college in Maine for a while, but in 1922 we married. We were married at St. John's Episcopal Church in Worcester on June 21, 1924. It was a noontime ceremony, and we had no attendants. I still wore a wedding gown, which was made of white georgette crepe over Canton crepe with a corsage bouquet. For our honeymoon, we took an auto trip through the White Mountains in New Hampshire. It was wonderful traveling through the beautiful countryside. It is a cherished memory.

On our return, we moved into the house at Maple Hill that he inherited from his grandmother. It needed a complete overhaul. I had all the new utilities and new windows installed, and I created a large living room by removing a partition. I had a fireplace, and a wraparound porch added to make the old house modern and comfortable.

Next, I began helping Robert get organized. I told one of my friends, "Most of the time he didn't know whether he had any money in the bank, or in his pockets. Fifteen minutes after dinner, he couldn't tell me what he had eaten. And some Clark students told me they saw him walking along a corridor, still holding an open umbrella over his head!"

To develop his social and recreational life, I took him to campus parties, picnics and bridge parties. He hated bridge and spent most of the time sitting in a corner smoking his cigars while I played. It was my job to see that he was happy and effective in his work, so I continually encouraged him. There were many detractors who thought his work was a fantasy, but I tried to make him ignore these naysayers.

I could see he needed help with his work, too. I typed his notes and organized his study. We bought some cameras, and I took pictures of his experiments. I took his now famous picture of the first liquid propellant rocket flight in March 1926.

I still remember those days with all the positive and negative publicity we got from the press and elsewhere. This continued throughout most of his life, but, despite it all, we could get support from the Smithsonian, and, in the late twenties, Charles Lindbergh became interested in our work. He could get financial support from the Guggenheims and a few others.

The fear that one of our rockets might land in a populated area prompted authorities to ask us to leave Massachusetts. With Lindy's help, we settled in New Mexico and began setting up a shop and a launch site near Roswell. I was deeply interested in Robert's work and became more and more familiar with the details of his experiments. I also enjoyed calling him Doctor G, especially after the Buck Rogers comic strip portrayed him as Dr. Huer.

He did some launches at Roswell, but the Depression caused us to lose Guggenheim funding, forcing our return to Worcester. He once again took a teaching position at Clark.

However, my husband's health was deteriorating in Worcester, and his doctor recommended he return to Mescalero Ranch at Roswell. Fortunately, the Guggenheim Foundation had renewed its support, which allowed Robert to make the move.

Throughout the 1930s, we experienced successes and failures at Roswell, anxiously waiting each year to find out if the Guggenheim Foundation would continue to fund our work.

During this period, Robert began work on the K-series rockets, which were larger and reached very high altitudes. They had a more powerful engine, but many of these engines experienced chamber burn-through.

"How can we prevent the chamber problem?" I asked. "Do you think a different cooling scheme might help?"

"I've tied different schemes, but they proved too complicated.[1] That's when I decided on curtain cooling, which sprayed excess gasoline around the inside wall of the combustion chamber to cool the metal." [2]

"So, what do you plan to do?"

"I'm considering going back to a smaller engine design, which worked better."

<><><>

THEREFORE, FROM 1936 THROUGH 1938 my husband worked on a smaller design, his L-series, which was about 18 feet. He conducted thirty tests with these rockets.

In these tests, he experimented with multiple combustion chambers and nozzles. In November 1936, he flew the L-7 with multiple chambers, hoping to increase thrust without increasing the size of a single chamber. It had four combustion chambers, the first rocket with multiple chambers and nozzles. It reached a height of 200 feet, correcting its vertical path by using vanes in the exhaust until one chamber burned through. The flight demonstrated that a rocket with multiple combustion chambers could fly stably and be easily guided.

On March 26, 1937, he achieved an altitude of about 8,900 feet with L-13. We reduced the weight of this rocket by using thin-walled propellant tanks wound with high-tensile-strength wire. Unfortunately, he never again reached that height with one of his designs.

Robert worked hard on his rockets, and I thought it was affecting his health. In August 1938, I finally convinced him that a vacation was in order, and we sailed to France on the Normandie. His health seemed to improve on the trip allowing us to visit Switzerland and England before returning to New Mexico.

Through 1941, he worked on his P-series of rockets, which used turbopumps to produce higher propellant pressures, resulting in greater thrust. His turbopump system burned a small amount of gasoline and liquid oxygen in a separate chamber, which he called a gas generator, with the hot gas passing through a turbine to drive the propellant pumps.[3] The turbopumps eliminated the need for tank pressurization, which reduced the tank weight.

When Robert mentioned the need for turbopumps to Harry Guggenheim, Harry suggested he contact pump manufacturers for technical and financial support. None were interested, as the development costs were prohibitive. We were left on our own to design and test the small turbopumps. Two launches with this design ended in crashes after reaching an altitude of only a few hundred feet. The turbopumps worked well, and my husband was quite pleased. However, these tests were the most trying and disheartening phase of his research because of the number of engine failures.

The Guggenheim Foundation not only funded my husband's work, but they also funded several others, including Theodore von Karman at Cal Tech. Harry Guggenheim and the other researchers felt a joint effort would be more productive, so Harry called my husband. "Doctor, I'd like you to come east and meet with the other researchers. We need to work as a team if we're going to be successful."

"I'll be there," Robert told him.

At the meeting, my Doctor G described his work including both his successes and failures.

“How can we best advance the research?” they asked him.

“There are several components in need of improvement. You could work on those for me.”

“We should all work on the entire rocket, so we can share our ideas as the research progresses,” von Karman answered. “One may have a solution to another’s problem.”

Robert didn’t want to give away his secrets, so he never joined the team.

<><><>

BACK AT ROSWELL, I continued to support Robert and to provide a warm and friendly home for him and hospitality for out-of-town visitors like the Lindberghs, the Guggenheims and officials from the Army and Navy and others who were there to see his rockets. The Navy Department’s lack of interest in the military potential of rocketry frustrated him, and the Army didn’t give him any support, either.

We had a group of friends who enjoyed activities we enjoyed, like singing. I encouraged Robert to participate in these activities to give him some diversion from his obsession with rockets.

In the fall of 1941, I noticed a growing huskiness in his voice. It was the early signs of throat cancer. At this time, America entered World War II, and the military became interested in the potential of rockets for the war effort. In 1942, the Navy Department invited him to move to Annapolis, Maryland, where he could set up his machine shop and test stands.

“Robert, I don’t think you should go there. It won’t be good for your health,” I cautioned.

“It’s an opportunity, I’m going,” was his answer.

So, we settled at Tydings in a roomy gray stucco house overlooking the Chesapeake Bay. From our home, we saw America's involvement in the European war as many ships set out from Baltimore harbor for the war zone.

The Navy had him develop a variable thrust rocket for their Jet Assisted Take Off program. The JATO unit was flight-tested on a Navy PBY seaplane in the Severn River and did what it was supposed to do, provide shorter takeoffs. Each test had longer burn times, but on one flight the engine burst into flames, which engulfed the rear of the plane and forced the pilot to land. That was the last PBY flight with a JATO unit.

Robert spent the rest of the war trying to develop his pump-fed engines.

Meanwhile, I enrolled at nearby John Hopkins University in Baltimore to pursue my studies, which were interrupted when I left Bates College to marry Doctor G in 1924.

My husband heartily approved of me going back to college, and I brought my mother in from Worcester to care for him and the house while I was gone. I received a Bachelor of Arts Degree in liberal arts with honors, but with Robert seriously ill, I didn't attend the commencement exercises.

<><><>

IN THE SPRING OF 1945, Robert saw a captured German V-2 ballistic missile in the naval laboratory in Annapolis, Maryland, where he had been working under contract. After a thorough inspection, he concluded that the Germans had stolen his work. The basic design of the V-2 was like his rockets.

Three features developed by my husband appeared in the V-2: turbopumps were used to inject the propellants into the combustion chamber under high pressure, gyroscopically controlled vanes were used to stabilize the rocket, and excess alcohol was fed around the inside of the combustion chamber to cool the internal walls.[4]

Toward the end of the war, he served as a consultant to the Curtiss-Wright Corporation, where he led the development of an engine, which became known as the XLR-25. This engine powered the Bell X-2 rocket plane, which flew at 2,100 miles per hour to an altitude of 126,000 feet.

My Robert died in Baltimore on August 10, 1945, the day before the Japanese unofficially surrendered, ending World War II.

After my husband's death, I ensured Robert's work was not forgotten and began going through his papers. I found some of his inventions had been stolen. I was angry, and I was determined to do something about it. First, I set out to get the patent rights to insure legal recognition for his work. I contacted Charles Hawley, his patent attorney, who said, "I think many of your husband's inventions haven't been patented. We need to look into the matter."

<><><>

WORKING WITH MR. HAWLEY, I could get 131 of his patents secured after Robert died for 214 in his name. Hawley said, "Many of Robert's patents are being infringed upon by the government. We need to do something about it."

With the Guggenheim Foundation, we filed a lawsuit. After about nine years, in 1960, we received a $1 million from the armed services and NASA, with the Guggenheim Foundation receiving half the settlement.

On September 16, 1959, the 86th Congress authorized the issuance of a gold medal in honor of my husband, and I was present for the formal dedication of NASA's Goddard Space Flight Center on March 16, 1961, 35 years to the day after he launched the first liquid-fueled rocket from his Aunt Effie's farm at Auburn, Massachusetts.

In 1963, Wernher von Braun, reflecting on the history of rocketry, said of my husband, "His rockets may have been rather crude by present-day standards, but they blazed the trail and incorporated many features used in our most modern rockets and space vehicles." He also said, "Goddard's experiments in liquid fuel saved us years of work and enabled us to perfect the V-2 years before it would have been possible."

I continued to promote Robert's accomplishments. In 1965, I donated his Magnesium Powder Experiment Box[5] to the National Air and Space Museum at Washington, D.C. This device was developed in 1916 and is the oldest space-oriented artifact in the museum and possibly in the world.

In 1976, he was inducted into the International Space Hall of Fame at the New Mexico Museum of Space History at Alamogordo.

Epilogue

ESTHER DIED AT HER HOME at Worcester, Massachusetts, on June 4, 1982, at 81. She was buried next to her husband at Hope Cemetery in Worcester.

Robert Hutchings Goddard
(1882-1945)
(Photo from the Public Domain)

Part Three

Hermann Oberth

Chapter 8 - Another Visionary

Our educational system is like an automobile, which has strong rear lights, brightly illuminating the past. But looking forward, things are barely discernible. – Hermann Oberth

Transylvania, 1917

DURING THE GREAT WAR, I was a doctor. After attending to my wounded and sick soldiers, while my comrades slept, I secretly worked on my rocket designs. I also calculated the amount of alcohol and oxygen needed to launch them.

One night, I didn't take my usual precautions, and I let a small amount of light shine under my door into the hall. This was discovered by the regimental physician, who entered my room demanding, "What's that?" pointing to the drawings and equations on my papers.

"Just a drawing, Herr Regimental Doctor," I answered nervously.

"Herr Oberth, what is this funny thing?" he demanded as he picked up my drawing of a rocket with a pointed nose and tail fins.

"Please, Herr Regimental Doctor, it's top secret."

Doctor Otto Csallner immediately responded, "If anything is top secret, I know of it. I saw nothing like this."

I didn't know what to say, so I remained silent.

"Well, I'll keep your secret. Now explain this drawing of a big skyrocket!"

"Herr Regimental Doctor, it is a rocket for the sky but not a skyrocket. "I didn't make this from cardboard, nor did I use gunpowder," I replied.

"Well, just what exactly is it?"

I explained it was a long-range missile made of metal, which I had invented. It was about twenty-five meters long and approximately five and a half meters in diameter.

The doctor fell into a chair and shook his head in disbelief. I knew he thought I had been working too hard and had lost my mind. "What makes the thing go? Wind, no doubt," he said sarcastically.

I explained the operation of my rocket, being careful to keep it simple, yet referring to the basic laws of physics.

"Never heard of such a thing," the doctor replied.

"It's something new. I've been inventing it since I was twelve, eleven years ago."

Dr. Csallner studied me and realized I was sincere, then asked, "Do you think this thing will really fly?"

"If my calculations are correct, it will travel hundreds of kilometers."

"What's it good for?"

"Sir, one or more sent to England would frighten King George into surrendering, and then the war would be over," I answered.

Dr. Csallner mulled over my invention and said, "If what you say is true, we must send your drawing to Berlin. Only Headquarters has the scientists to evaluate it."

"Will it be safe in the ordinary mail?" I asked.

"I'll find a way. In the meantime, show it to nobody!" He then left without saying another word to me.

<><><>

NAMED HERMANN JULIUS OBERTH, I was born at Hermannstadt, Austria-Hungary[1] on 25 June 1894. My father, Dr. Julius Oberth, was a highly respected doctor in our little hamlet in the Transylvania section of our country. I can still remember talking to my father about the Moon, when I was about six. As he drove up to our house, I ran to him and asked, "What is the Moon made of"?

"It's a big stone ball with mountains," he replied.

"Can we go there?" I asked.

"Maybe someday someone will invent a Moon ship. You may even be the one to do it," Father answered.

"Do you think I could?"

His answer was profound, "A fellow can do anything, if he wants to enough."

I never forgot his words, especially later in life, when I met severe opposition to my ideas.

When I was about eleven years old, Mother gave me a book written by Jules Verne, *From the Earth to the Moon*. The story was about a nine-hundred-foot cannon buried in the ground, which fired a nine-foot diameter ball to the Moon from a place called

Florida. Influenced by Verne, I began talking about going to the Moon and asking questions. The answers I got were, "Forget the Moon, you can't go there," and "Your notions are against tradition and therefore wrong."

From these comments, I made a resolution: In the future, before I give up my opinions someone must actually prove them wrong.

In the autumn of 1913, I went to Munich to study medicine. My father was very pleased and expressed hopes of me taking over his place at the hospital. I also took some science courses I thought would help me in designing a rocket someday. My studies didn't last very long as the Great War broke out the following summer. I enlisted, and they assigned me to an infantry battalion, and sent me to the Eastern front against Russia. In 1915, they moved me to a medical unit at a hospital in Sighisoara, Transylvania. There, I found some time to work on rocket designs.

<><><>

I HOPED THE GERMAN HEADQUARTERS would accept my rocket design, which would help me to convince Father that I should be a scientist and not a physician.

Meanwhile, Dr. Csallner had become convinced that my rocket was most likely the only way we could win the war against Great Britain. However, he realized it would be difficult to send my it to Berlin through the normal chain of command of subordinate officers. Therefore, he told me to take my plans directly to the German Consul and, "Leave as soon as you can get ready!"

I took my work to the Consul in Kronstadt, just a few miles away, and asked him to send it on to Berlin. He didn't seem to understand it and was skeptical, but he agreed to send it with his own papers. "It will take two weeks for a reply," he told me.

I returned to the military hospital at Schaessburg and awaited the news from Berlin. In the meantime, I once again visited Mathilde

Hummel, who I had met years before. She was a blue-eyed beauty, whom I affectionately called Tilly. I explained my rocket ideas to her, and she was genuinely interested. We dated more and more, and I hoped to tell her that Berlin had accepted my plans for a rocket to end the war.

It was not to be. Finally, I received an official letter from Berlin's Crown Council. It read, "We have carefully studied your sketches. Surely, you know that rockets travel no farther than seven kilometers, not hundreds. We are certain your calculations are wrong. Therefore, we must reject your plan." My original documents were also returned with the letter.

I dreaded telling Tilly and her two sisters, who all believed in me and in my rocket. I finally mustered the courage to give her the news from Berlin.

"Don't give up!" Tilly said. "Someday, the world will listen to you."

Tilly didn't give up on me either. She agreed to my proposal, and on 8 July 1918, we were married. After a short honeymoon, I returned to my post at Schaessburg.

That November, the war ended, and they discharged me from service. After spending only a short time teaching at the University of Siebenburgen, before the university closed, I returned home to Schaessburg. I was now the father of Julius, and I was out of work.

Fortunately, Father agreed to me studying mathematics and physics, instead of medicine, and he supported me financially while I pursued these studies. I attended the University of Munich, but there was no place where Tilly and I could live, so she remained in Schaessburg. To find an apartment for the both of us, I transferred to Gottingen University in March 1920. Tilly helped us find an apartment, but landlords repeatedly rejected us because we were foreigners.Disappointed, Tilly returned to Schaessburg.

By this time, I was prepared to pursue a doctorate. I thought a dissertation on rocketry would be worthwhile. I approached Dr. Ambronn, a professor at the university, with the idea, but he was unsupportive.

Next, I went to Dr. Wiechert, the inventor of an early seismograph, but he simply folded my papers and said, "I can't go aboard with you on this outer-space stuff. We will never escape the atmosphere."

Undaunted, I went to a teacher of aerodynamics, who concluded, "Too many mistakes."

I rechecked my calculations and found no error.

By the spring of 1921, I had come up with another design, a double rocket, with the two sections of the rocket mounted in single file where the lower section would fall away after the fuel burned out. Fuel for the first part would be alcohol and for the second part hydrogen. Both sections would have oxygen for the combustion process. My calculations indicated if there is a small rocket on top of a big one, and if the big one is jettisoned and the small one is ignited, then their speeds are added. The design appeared workable, but only testing would prove it.

I worked that summer at the university, but in the fall transferred to the University of Heidelberg. Again, we couldn't find an apartment, this time because we had two small children. Tilly returned home once again, and I worked on another rocket design and pursued my doctorate at Heidelberg.

In 1922, I submitted my thesis to the university, but the university rejected it. I wrote to Tilly, "I will not write another doctorate dissertation, I will prove that I can become a greater scientist than some of these professors, even without the title of doctor."

Later, I told her, and others, "Our system of education is like an automobile, which has strong rear lights, brightly illuminating the past. But looking forward, things are barely discernible."

I sought an expert opinion about my paper from the astronomer, Dr. Max Wolf.

"Congratulations, young man. An excellent paper. Publish it," was his response.

I sent it to a publisher and waited for a reply. Finally it came, "We dare not publish anything so fantastic as a book about a rocket and space travel lest we lose our name among scientists."

About the same time, I read about Dr. Robert H. Goddard's work on rockets. He was the head of the physics department at Clark University in Massachusetts. I asked him to work with me to develop a rocket. He, too, turned me away.

Not wanting to admit failure to Tilly, I secretly sent the manuscript to another publisher, then another and another. I submitted it six times, and each time it was rejected. Finally, I told Tilly.

At first, she was silent, then she said, "You can succeed without a doctorate."

She surprised me with her next statement, "Publish it yourself! I have saved ten thousand lei,[2] Use that!"

After much searching, I found a company willing to publish it. To minimize the contents, the company would only publish my book if there was no picture on the cover, only the title and author. In June 1923, *By Rocket into Planetary Space*[3] went into print and the publisher and Tilly and I were completely unprepared for what happened next.

Cover of an early English translation of Jules Verne's book, circa 1870. (From the Public Domain)

Chapter 9 - The Girl in the Moon

The rockets... can be built so powerfully that they could be capable of carrying a man aloft. – Hermann Oberth

Schaessburg, 1923

I TRIED TO MAKE MY BOOK the result of scientific observations and calculations. I wished to prove it is possible to build a machine to go beyond the Earth's atmosphere. These machines would reach enough speed so that they would not fall back to Earth. They could be built to carry men, and someday, when the economy improves, maybe several decades in the future, these machines could be built.

The book was divided into three parts. Part I was intended for engineers, astronomers and scientists. It contained many equations, which described the basic principles of rocketry. It also raised the possibility of using the rocket's own fuel passing through tubes on the outside of the chamber to cool the inner walls.[1]

Part II described my Model B rocket, which I had designed earlier. It was about five meters and had a lightweight metal skin. I finished this section by writing, "I consider in rocketry up to now there are two new essentials: the use of liquid fuel instead of the common explosives, and a two-stage or divided rocket. Both I have advanced."

Part III is where I looked way into the future. My Model E rocket was huge, capable of carrying men into space. I was concerned about weightlessness, but concluded humans could adapt to it, although there may be some fright or something like seasickness at first. Takeoffs and landings might be dangerous, but we could manage them, especially with a water landing in an unobstructed area.

Meteoroids are also a danger. However, large ones that could penetrate the rocket's skin are extremely rare.[2] Men would need to wear spacesuits with hooks to grasp rungs on the spaceship's walls, and they would need oxygen and nitrogen tanks to replenish the air supply.

The spaceship could carry instruments to study radiation and light waves. It could circle the Moon and study the unknown side. Certain rockets could remain in space and become little moons capable of relaying straight-wave radio signals to different parts of the Earth.

One particularly interesting concept is to use a little moon to monitor icebergs. If such a device had been in place in 1912, they might have avoided losing the Titanic. Another concept I described was a space mirror. It could direct the sun's light to melt icebergs and frost to save crops or control the weather by destroying tornadoes or creating depressions to induce rain.[3] One use, which is not so positive, is a space weapon to blowup munitions factories and depots, set enemy towns ablaze and to scorch troops.[4] [5]

Finally, I added, "I do not wish to close this book without saying that it is possible to build all the foregoing." But I will not claim that it will happen in the next ten years."

My book had a mixed reaction. Professors, physicists, and astronomers attacked my discussions on space. A Privy Councilor wrote in the *Journal of German Engineers*, "We believe that the time has not come for delving into such problems as these and probably never will come."

A professor wrote in the review magazine, *Die Umschau*, "The author clearly has only a limited knowledge of equipment required for space experiments. . . .As to the possibility of taking this apparatus into space where there is no gravity and no air, it can only be said that the reactive driving power of a rocket depends on the presence of a mass of air."

I asked for space in numerous magazines in order to respond to my critics. After many refusals, *Die Umschau* gave me space. I wrote in the March 22, 1924 issue, "Of course there will be technical problems yet to solve in building a large rocket. But in the seventeen years I have been working on them, I have found nothing to prove that rocket manufacture is impossible or even doubtful. It can be done."

The general public loved my book. It quickly sold out and caused me to publish a second edition. A young man named Wernher von Braun wrote he could not understand all the equations.

However, I never received compensation for building a rocket. I continued to work teaching high school mathematics and physics at Mediasch to support my family, as there was virtually no work available at the college or university level in these hard times.

Because I received a negative reaction from the scientific community, and such a positive one from the general reader, I wrote another book directed at the population to make my ideas popular. Thus, I wrote *The Road to Space Travel.*[6]

<><><>

BY 1927, PEOPLE BECAME ENAMORED with space travel, probably because of the poor German economy. Germany was required by the Treaty of Versailles to pay war reparations, which took its toll on the country and the average citizen. This created an atmosphere of escapism, which precipitated this interest.

Because of this diversion and hopefully my book, a group of enthusiasts formed what eventually became known as the German Rocket Society. Within a year, this group grew to five hundred members, with me being one of them.

With all the space enthusiasm, the German company, Universal Film Corporation, cashed in on the excitement. They had Fritz Lang, a famous director, lead the production, *The Girl in the Moon*,[7] which ensured it to be a success. It was based on my book, and I was invited to be the technical advisor for the film.

I was excited, to say the least. I told Tilley, “Maybe the movie would generate enough interest so the German government would fund me to build a rocket.”

I took a leave of absence from the high school and went to Berlin to meet Mr. Lang. At his home, he described the script of the film.

The story about six men landing on the Moon shocked me. ”Herr Lang, there’s no air on the Moon, so how ----?”

“We’ll put some there.”

I was worried about this film. With my name attached to it, it could ruin my reputation as a scientist.

Before I could say anything, Mr. Lang delved into more of the film by showing me sketches of the surface of the Moon to be included as backdrops in the film. He motioned for me to be silent, then said, “Don’t complain, Professor, we can do anything with a poetic license.”

I left Lang’s house with mixed emotions. Obviously, he was not interested in scientific accuracy, just in having my name attached to the film for publicity. The film company promised me a workshop, laboratory and helpers to build a small rocket for the movie. Willy Ley, a famous writer of stories on space travel, had urged Universal Film to build a real rocket and fire it for publicity's sake.

The German Rocket Society was glad to have a now somewhat famous rocket expert in their organization, so they elected me president. Encouraged, I could obtain another leave of absence from the school to build a rocket, which Universal Film now endorsed. I was using liquid oxygen and gasoline as propellants, not the least concerned these were very hazardous fluids to be working with.

About this time, the spring of 1929, I received notice that I was being awarded the first REP-Hirsch Prize for my book. In fact, the sponsors of the prize said they had doubled the prize money to ten thousand francs[8] because they were so impressed with my book.

I immediately called my friend Willy Ley and asked, "Is this true?"

"It certainly is. The French think you have done the most for space travel, which is why you won and received a larger sum."

This gave me the resources to continue building my rocket, but I ran into a major problem when the film company announced the movie premier would be October 15th, 1929. I didn't think I would be ready by then. I needed more time and more help. More time was not available, but Willy Ley and I recruited additional workers to manufacture the many parts of my rocket. We accepted anyone we were so in need of help.

Willy recruited a lad of about seventeen who dreamed of flying to Mars and Venus. He introduced the boy to me.

"Herr Professor, though I'm still in school, I'd like to work for you in my spare hours."

I accepted the boy. His name was Wernher von Braun, who later became a major figure in my life.

Everyone was happy with the film, except me. I didn't finish the rocket for the premier, a small model rocket was used instead. My real rocket was no longer of any use to the company, and I was very disappointed that I had never could test the liquid propellants. When asked what to do with it, I said, "Sell it, and use the money to build a bigger one." I was with no funding and thought myself a failure. Thus, I returned to the classroom at Mediasch.

<><><>

THE REP-HIRSCH PRIZE brought me more publicity. Once again, I was attacked by the so-called experts. They said meteoroids, the sun's rays or even the Northern Lights would kill a passenger in a spaceship. One went so far as to say, "Assaulting the gates of Heaven by going into space is sacrilegious, and, therefore, it's certain suicide."

I didn't bother to answer these unproven accusations.

I finished the school year with the desire to try again to get some support to build a rocket. By now, we were in the Depression, and money was scarce. Rudolf Nebel, who had worked for me on the film project, went with me soliciting funds. We faced repeated rejection, but Dr. Ritter, the Director of the *Chemisch-Technische Reichsanstadt*,[9] proved a willing listener.

After looking at our drawings, he told us, "I have no money to support your research, but if you can demonstrate the liquid propellant rocket motor, I can give you a certificate to help you solicit funds."

I was delighted to receive this amount of encouragement. I used Cone-nozzle,[10] the nickname the crew gave to the real rocket I was working on for the movie.

When the German Rocket Society heard the news, three members volunteered to help me: Wernher von Braun, Rudolf Nebel and

Klaus Riedel, an engineer. We conducted the test in the woods outside of Spandau, a suburb of Berlin. Dr. Ritter was present.

The propellants were loaded and introduced into the oven,[11] and the motor ignited with a burning rag on the end of a long pole. The motor started with a deafening roar. It didn't fail but shut down when the propellants were spent.

Later, Dr. Ritter wrote a letter concerning the test, "The motor had performed without mishap on July 23, 1930, for 90 seconds, consuming 6 kilograms of liquid oxygen and 1 kilogram of gasoline, and delivering a constant thrust of about 7 kilograms." [12]

The certificate from Dr. Ritter was presented to the German Rocket Society, who then planned a permanent laboratory with all necessary rocket building and testing equipment. They also looked for a field where a rocket airdrome could test the rockets.

I wrote to Tilly, "I have good news. The German Rocket Society is going to support my research."

But my euphoria was short-lived. I received a letter from the school saying they could not extend my leave of absence. Either I return, or my position would be filled. I had no choice. With a wife and four children, I taught. I still dreamed about rockets and hoped I would see in the newspapers the German Rocket Society had done something significant. All I saw was the Nazis had won the election, and Adolph Hitler was gaining power in Germany.

<><><>

ON A SNOWY AFTERNOON, someone came to my door. He didn't tell me who he was, and with discord between Germany and Rumania, I didn't ask, thinking he was from the German Rocket Society. However, he wasn't German, although he spoke in broken German. This shabbily dressed man ordered me to close the window shades. When I went to my desk, he came near to me and said in a very low voice, "Herr Professor, I come to talk to you on an important matter. I'm sure you will not reveal my visit. You are wise enough to know talk is dangerous."

Hermann Oberth (center, in profile) demonstrated his tiny liquid-fuel rocket engine in Berlin in 1930. Second from the right is 18-year-old student Wernher von Braun. (Photo from National Air and Space Museum, Smithsonian Institution, Photo Number: NASM 2003-3754)

Chapter 10 - The V-2

If there is a small rocket on top of a big one, and if the big one is jettisoned and the small one is ignited, then their speeds are added.
– Hermann Oberth

Rumania, 1932

THE MYSTERIOUS GUEST sat down next to my desk and whispered in my ear, "Professor Oberth, I come from the Soviet Union to ask if you are ready to work for us. We know all about you. We have your drawings."

I was incredulous. I didn't know whether to believe him. He may be from the Rumanian security police to test me. I quickly tested him instead. "Let me see your passport or some identification."

"I could go to jail with a Russian passport in Rumania. I came across the border illegally, but I have the best identification possible." With that, he pulled from his pocket a big roll of money. "There! Ten thousand lei to start and three thousand rubles a month plus everything you need: house, laboratory, and equipment, all tax-free. What do you say, Professor?"

This was very tempting, but I was suspicious. "What would I work on for the Soviets?"

"On what you like most, rockets."

"As war weapons?"

"We don't want war, but if we need to defend ourselves against capitalists, that's a different situation. We would be forced to use rockets if they attacked us."

"Do the Soviets know how to build rockets?" I asked.

"Show me one of your drawings."

I pulled an incomplete drawing from my desk drawer. It was a sketch of a steering device I was having trouble designing. He studied it, took a pencil and added a lever in just the right place. I was amazed. This man solved my problem and was obviously well versed in rocketry. The Soviets were very advanced in rocket design.

"Now, Professor, what do you say?"

"No! Thank you," I replied.

The man destroyed the sketch and picked up the lei. "I'll be back. Remember, you're a father. We have eyes and ears everywhere. Yes, everywhere." With those remarks, he left as quickly as he had appeared.

I sent a letter to Berlin informing them of my encounter with the Soviet agent. They quickly responded and asked me not to work for Russia. They didn't ask me to work for them either. At that point, I did not know what was going on in Germany regarding rocketry.

In some magazines, I found the German Rocket Society had acquired a workshop and test site. Willy Ley, a writer, with the help of an engineer, built a rocket. But the Society was short-lived. Hitler became *der Fuehrer* with complete control of Germany. All rocket work was to be conducted for the "national defense." As a result, the German Rocket Society dwindled, and Willey Ley prepared to go to America. The Army Ordnance took control of German rocketry. Those wanting to work on rockets joined this group at an experimental station near Berlin run by Captain Walter Dornberger. I stayed at Mediasch, where I worked on what I pleased.

<><><>

MEANWHILE, GERMANY WAS BECOMING a police state, with secret police arresting people for not conforming to Nazi expectations, such as failing to give the "Heil Hitler" salute. We weren't able to eat much besides bread and vegetables, as all "surplus" food went to der Fuehrer. My oldest daughter, almost fourteen, would soon join the Hitler Maidens and do a half year's work for nothing. Little Ilse would be trained in the Young Maidens, and our son, Julius, would be enrolled in the Hitler Youth. I had seen little of my children, being away so much. Now, they were being trained by the Nazis. "Is all this necessary?" I asked Tilley.

"I don't know. Everything is happening so fast, and I'm worried."

In Germany, there was a buildup of a "defense force" supposedly to stem the Red Tide from taking over lands outside of Russia. Other countries were also building up their defenses, and I received an offer from Japan to build rockets for them. I quickly refused.

A third offer came from the King of Rumania to come to his palace. "I summoned you here to discuss rockets. Professor, will you build a rocket institution for Rumania? You will have full government financing."

Being a Saxon, Germany was where my loyalty lived. I quickly answered, "Forgive me, Your Majesty, I have no talent for organization. Permit me to continue to work in Mediasch."

The king didn't say another word. Instead, he called for an escort to take me to the railroad station for me to return home. As with

Russia and Japan, I wrote to Berlin telling them about Rumania wanting to develop rocket technology.

About this time, I learned Major Dornberger had built a rocket with 270 kilograms of thrust and was going to build one with about 900 kilograms of thrust.[1] They were looking for a field away from the populated areas where they could conduct tests and finally settled on Peenemünde, an island in the Baltic Sea.

I thought about these events. Russia, Japan and Rumania wanted me to help them build rockets, but my fatherland had requested no help from me in their work at Peenemünde, despite my letters on how foreign governments were working to develop rocket capability. Once, a couple of German officers visited me and asked me about my work. "Thank you very much, Herr Professor. You'll hear from us." That was 1935.

I heard nothing for months. I felt like a failure once again. Then, in the spring of 1937, I received a letter from the Reich Air Ministry. I was to be included in developing rockets for Germany.

Tilly and I agreed I should go to Berlin, where I would work, and look for housing for her and the children.

At Berlin, I met with Major Dornberger and Dr. Werhner von Braun and others. Over drawings and documents, some of which I had written, they questioned me for hours.

At the end of our discussions, Major Dornberger gathered up the plans and told the Professor, "Stay where I can reach you."

"But I came here to work on rockets."

"Not yet. We'll let you know. Good day."

After a while, I returned to Mediasch, again feeling like a failure. I was good enough to answer their questions, but not good enough to work with them.

A year later, I received a letter asking me to join a rocket program at the Technical College in Vienna. At least this was a part of the Greater Germany. I accepted, only to learn I was to substitute for another professor who was working on more important things.

I delved into my research, anyway. Before long, I was transferred to the Technical College at Dresden. I protested but was told I was to research the propellant pumps for a new rocket. I perfected a pump design, submitted it to the authorities, only to be informed that an engineer at Peenemünde had designed one like it. I was angry and told the top officials at Dresden, "I'm through. I'm going back to Mediasch."

"Herr Professor, that is forbidden for you. You know too much to leave Germany."

"Then send me to Peenemünde."

"Impossible! That's top-secret, and you're a Rumanian."

I pulled my paperwork and showed the officials I was a German citizen.

I was told, "We'll call you."

Tilly joined me at Dresden, which made my life pleasant, despite the minor role I played in the German rocket program.

On 1 September 1939, we were told, "Poland attacked the Reich." Shortly after, England and France declared war on Germany. The newspapers predicted a quick war, but it soon spread over most of Europe. My son, Julius, was called into the army. Then, in August 1940, the air war over England began.

<><><>

FINALLY, A LETTER ARRIVED. Peenemünde could use me. My dream was coming true. I took a train to the base and was met at the gate by Dr. Wernher von Braun. We got into a sedan and drove along the paved streets. I learned he was working on the A-4 rocket, an abbreviation for Aggregate Number Four.

The base was huge. There were seven thousand engineers, scientists and technicians working to develop the A-4. The finest electrical, structural and hydraulic labs in Germany, plus metal and woodworking shops, were there. I later learned it even had a supersonic wind tunnel, which was the world's first. To keep everything top-secret, barbed wire fences and armed guards surrounded the entire operation.

"My friend," von Braun said, "we may lose all of this if the A-4 fails."

"Why is that?" I asked, totally surprised by his statement.

"Hitler is not a lover of rocketry since a friend of his was killed in a rocket accident years ago. He may shut down the work at Peenemünde and use the personnel and equipment here for other parts of the war effort. To save the A-4 project, we need a successful flight test."

I was given a clearance to visit various sections of the base. I was no longer Hermann Oberth; I was now Fritz Hann to keep my presence at the base as secret as possible. In the ensuing days, there were several ground tests and flight test of the A-4. The ground tests were failures, but more importantly, the flight test failed, too. I found the rocket was based on my original drawings and didn't have the improvements I had developed over the years.

When I offered to help, I was told, "Hitler's war won't wait. The A-4 now goes into mass production."

About that time, it looked like the war in Russia would end in a German defeat and there would be a revolt on the home front. To counter this, the Ministry of Public Enlightenment and Culture turned to the A-4 as a solution. They praised the A-4 as a new weapon destined to completely change the art of war. It was renamed the V-2, the V standing for *Vergeltungswaffe*, or Retaliation Weapon. The inner circle at Peenemünde knew the rocket was unreliable and needed to do something. Thus, work intensified to make the V-2 a success.

I was doing routine work and not involved in this effort, when one day, in desperation, Dr. von Braun came to me and asked, "Herr Professor, can you develop an Atlantic rocket to reach New York?" I nodded, and he continued, "We must get to it right away. No time to lose. Show how it will look, what will be the cost, and how much fuel will be needed. Let me know as soon as possible."

I designed a three-stage rocket. The two lower stages were to burn first and fall into the ocean with the upper stage reaching New York. It would carry a payload of one metric ton.[2] I called Dr. von Braun to show him the drawings, but before he could meet me, the Japanese attacked Pearl Harbor in Hawaii.

Now, with the United States entering the war, von Braun knew all efforts must be directed to making the V-2 a success to save the program and help Germany win the war. He put aside my drawings and everyone concentrated on another V-2 flight test as soon as possible, which took place in June 1942. This test failed in flight, at about an altitude of five and one-half miles, when the nose cone came loose, causing the rocket to become unbalanced and to break up. Dr. Dornberger was very disappointed but at the same time determined to make the V-2 a success. After a scramble, to correct the deficiencies, another test was conducted in August only to fail again.

At this most inopportune time, the air force came out with a competitor before we could make changes and test again. The V-1 rocket, or "buzz bomb" as the British came to call it entered service. Fortunately for our program, it had limited success. It wasn't a true rocket, as it used air for the oxidizer, and it was slow, flying at only 580 kilometers per hour.[3] Enemy anti-aircraft guns and planes easily shot it down. In fact, as planes became faster, pilots would touch its wing with their plane's wing and send it in a direction away from England. Our next test must succeed; otherwise, the V-2 and Peenemünde projects are over.

I'll never forget that day in 1942 when we fired the first successful V-2. The 14-meter long, 12,500-kilogram rocket[4] roared into the sky, propelled by more than 22,700 kilograms of thrust.[5] It reached a height of approximately 88 kilometers[6] and a speed of 5,800 kilometers per hour.[7] After the flight, Dr. Dornberger gave a speech and said, "This third day of October 1942, is the first of a new era in transportation, that of space travel..."

During the celebration that night, it was pointed out that the V-2 had proven that: (1) It is possible to fly with an automatic guidance system, (2) It could be controlled at the speed of 5,800 kilometers per hour, (3) A rocket could operate in an airless environment. Now, the German High Command would send funds for further development. In addition, the room was filled with all kinds of speculation, such as quick travel between continents and larger rockets going to the planets.

As I left the building, I congratulated Dr. Dornberger.

To my surprise, he said, "But, Professor, congratulations go to you for showing us the way."

I was astonished and hardly knew what to say, not having worked on it very much. But the euphoria of everyone there, and mine as well, were short-lived. The Allies had other plans for Peenemünde.

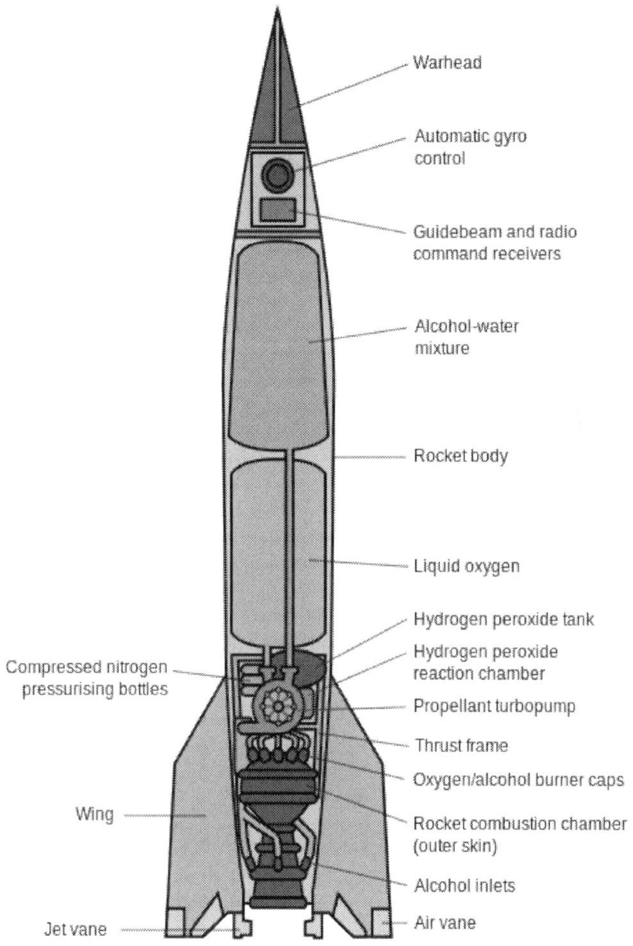

Schematic of a V-2 rocket.
(from NASA files in the Public Domain)

Chapter 11 - Man into Space

It is possible that men will one day reach distant, unknown planets of other suns. – Hermann Oberth

Peenemünde, 1942

ALTHOUGH THE V-2 FIRING on 3 October 1942 was successful, it took another two years to get it into production. Hitler had a dream that no V-2 would ever reach England, and he took it as fact. Not getting the support needed to finish development of the rocket, Colonel Dornberger and Dr. von Braun repeatedly went to the High Command and to Hitler himself to get the necessary assistance for the project. When Hitler finally gave the V-2 top priority, many of the scientists thought it was already too late to save the Reich.

Meanwhile, in February 1943, a German general surrendered two hundred thousand troops to the Russians. Like Napoleon's attack on Russia more than a century before, the harsh winter led to the German defeat. However, I was more worried about my son, Julius, who was serving on the Russian front and may have been among the troops captured by the Russians. The Russians had a reputation for their harsh treatment of prisoners.

Tilly, in her letters, shared my concerns about Julius. She also wrote about the hardships at home. "Our houses have no heat, and to keep from freezing, I wear several sweaters. Also, food is being rationed. They give us only a small amount of potatoes and bread."

I knew they sent anyone who complained to a labor camp, so I wrote, "Things will eventually improve."

In June, Hitler demanded full-scale production of the V-2 by October 1943, so an all-out attack on the Allies could begin then. However, in August, Allied reconnaissance discovered the factory at Peenemünde.

Fortunately, I hadn't brought Tilly to the rocket base, so when the Allies bombed it on August 17th, I had only to look after myself during the raid. There was considerable damage and several hundred fatalities. But the wreckage and loss of life were not enough to prevent the continuation of V-2 development. The leaders were ingenious in their reaction to the attack. To make the base appear severely damaged, they dug craters and added camouflage to key areas. Despite the destruction, the High Command still wanted thirty V-2s a day.

I worked hard to rebuild the base, knowing no one abandoned working for the German Army. Several weeks after the initial attack, I received the Cross of Merit with Swords for my work in reconstructing the site.

By December, it was apparent the repeated Allied attacks throughout Germany were preparation for an invasion of Europe. They brought more forced laborers to Peenemünde to produce V-2s as quickly as possible.

Also, realizing an Allied invasion was imminent, the High Command developed an anti-aircraft missile. One leader at Peenemünde called for me, "Professor Oberth, you've discussed the possibility of a solid-propellant rocket for anti-aircraft defense purposes. Therefore, we are assigning you to the Waterfall[1] program to create one for the Reich. Because Peenemünde lacks the facilities to develop solid propellants, we are moving the project to Wittenberg. You will begin work there next week."

I had mixed feelings about leaving Peenemünde. I felt once again I wasn't being included in the major effort, but I felt I might save the Third Reich with this new weapon. Although developing such a system so fast was a large order, it was desperately needed. The Allies had landed at Normandy in June 1944 and were moving toward Germany.

It wasn't until September 1944 that the V-2 was used against England and Western Europe. I later learned, more than 3,000 V-2s were fired, killing about 9,000 citizens and military personnel, while about 12,000 prisoners of war died producing the weapon. Meanwhile, I was close to developing a rocket, which used ammonium nitrate, but by that time the chemical was no longer available, as it was extensively used for other munitions.

Then bad news came in a letter from Tilly. "Ilse died in a terrible explosion." Our daughter had taken a position as a chemist at a liquid oxygen factory to allow a man to fight in the war. Julius was missing and now Adolph, only fourteen, was called to serve in an anti-aircraft unit.

By 1945, it was evident Germany's surrender was imminent. The Russians were coming into Germany from the east, and the Allies were closing in from the west. The operation at Peenemünde was moved to the caves at Nordhausen before Peenemünde fell to the Russians. Soon after, the Allies captured Nordhausen, and they shipped several hundred boxcars of V-2s to the United States. My capture was certain. I knew I couldn't make it back to Tilly in Feucht. It was just a matter of who I would rather be captured by the Americans or the Russians. I chose the Americans, because I knew I would get better treatment from them.

After Germany surrendered in May, I made my way through the ruins of the Third Reich to Bavaria. At Regensburg, I was picked up by an American unit, and I was interrogated for hours about rockets. A few days later, I was sent to a town outside of Versailles in France. We were under armed guard and not permitted to talk to other prisoners, although we had all the physical comforts, we lacked prior to being captured. We were repeatedly interrogated, and I just wanted to be released.

Instead, I was sent to Kronberg, near the Rhine River. In the interrogation process, I gave them simple answers about my work at Peenemünde and Wittenberg, but the interrogations continued. I tired of the same questions repeatedly. I wanted to go home. So, I tried a different approach. I pretended not to remember, and I told the interrogator, "I'm sorry, sir. I guess I was not sufficiently interested in that detail to remember it." I was promptly released.

Once home, I tried to console Tilly. Ilse was dead, and Julius was still missing. On top of that, I couldn't find any work to support us. My only recourse was to cross into Switzerland and work there. I disguised myself as a mountain guide and gave the border guards the name I had used at Peenemünde, Fritz Hann. But work in Switzerland was not steady. I just provided technical advice and wrote magazine articles. Finally, my efforts to find suitable work paid off. The Italian Government wanted an anti-aircraft missile, as one of their submarine bases suffered severely from air attacks during the war, and they didn't want that to happen again.

At that time, the Italian border guards didn't allow Germans to enter Italy, so I took the pseudonym John Smith, and claimed to be an Englishman. Tilly later joined me under the name Mrs. Smith, and I hired three Italians and a chemist from WASAG[2], the company I worked with, to develop an anti-aircraft missile during the war. The chemist became Mr. Jones in order to enter Italy.

The contract lasted until the end of 1953. We had several successful launches, and the Italian navy seemed pleased with our work, although I heard nothing more about it after I left Italy.

Tilly and I returned to Feucht, where things were quite different. I felt very lonely and old, being just under sixty years of age. I said to Tilly, "I feel we are all alone. My parents are dead. Ilse is dead, and we never heard from Julius, so he is probably among those captured by the Russians, who died in one of their forced labor camps. Our daughter, Erna, is married and is a practicing attorney with a home of her own. Our son, Adolph, is a chemist with a life of his own, too. And our friends have all moved away."

"I know. I feel the same way. How different things are than before the war."

To keep myself busy and not to dwell on these things, I began writing a book on outer space. I gathered together my notes and pondered about exploring the Moon. I conceived a space suit, a pistol-like device for moving a man in space, which utilized Newton's law that for every action there is an equal and opposite reaction. In 1954, I published this work under the title *Man into Space*.[3] Several years later, I published it in America.

About the time *Man into Space* appeared in Europe, flying saucers were being reported in America. The United States Air Force classified them as Unidentified Flying Objects, or UFOs. The Soviets thought these sightings were a ruse by various organizations to pry money out of Congress for research work. As a result, American journalists turned to space experts to get their opinions on what the UFOs really were. I was one expert.

In the *American Weekly*,[4] I said, "It is my thesis that flying saucers are real, and they are spaceships from another solar system."

Some supported my thoughts, and others did not. There was a lot of interest in space. There was talk of a space station on an asteroid[5] for future space travelers. Some thought the asteroids were the remains of a planet that exploded in time past. I mentioned these in my book and thought about our future. I wrote, "People with imaginative minds even speculate whether this vanished planet was blown into the air (or rather, into space) by its own inhabitants, and this idea is put forward as a timely warning. Within the foreseeable future, the Earth's inhabitants will unquestionably have the means and the power of blowing their own planet to pieces.”[6]

<><><>

IN THE SPRING OF 1955, Dr. von Braun invited me to work on rockets in the United States. In July, Tilly and I arrived in New York, and we quickly went through the immigration process. We went to Huntsville, Alabama, to join the team of German scientists working under von Braun for the U.S. Army Ordnance Corps.

It wasn’t easy for the Germans in their new home. They and their families were sometimes called Nazis, which required extreme patience to avoid confrontations with the local citizens. Despite this, we learned English, became active in community affairs and eventually some became naturalized citizens.

America was interested in its satellite program, and I, too, caught the enthusiasm. I noticed that many of the rocket principles I had developed over thirty years ago were now being applied to the space program. I later wrote, “In fact, I’m rather surprised that so many of my former suggestions have been adopted in modern rocket development. Among others, only these should be mentioned: Regenerative cooling, reinforcing tanks by internal pressure,[7] liquid fuel such as alcohol containing water[8] and liquid oxygen, and finally the use of liquid hydrogen as fuel in upper stages.”

The space program validated my work regarding a rocket’s flight path. I had suggested what is now called a *synergic*, or combined curve, trajectory. Horizontal and vertical launches are almost equivalent in the propellants needed for a rocket to reach a given speed. The horizontal path takes advantage of the Earth’s rotational speed but encounters more aerodynamic drag because of its longer flight through the atmosphere. Also, the high drag requires a heavier structure with more reinforcement. To minimize the drag and take advantage of the Earth’s rotation, I suggested a vertical launch, and once outside the Earth's atmosphere, the trajectory is bent over to a horizontal one.

My work in Alabama was short-lived. The German Government would not send a pension check to anyone living in a foreign country, and I hadn't spent enough time in America to receive much of a pension there. My only recourse was to return to Feucht, Germany, where I retired and thought about many things.

In 1959, I published *Material and Life*.[9] In this work, I contended "materialism, the philosophy on which communism is based, is incorrect and further that aspects of human life, such as the soul, could not be explained by material reason."

I also thought about my life's work. To what purpose had I spent my life? I had given my answer in *Man into Space*: "To make available for life in every place where life is possible. To make inhabitable all worlds as yet uninhabitable, and all life purposeful."

Retirement had its rewards. The West German Government gave me its highest award to a scientist, the Great Cross of Merit, and the Iowa Wesleyan College gave me a degree. Now, I became Dr. Oberth.

Epilogue

HERMANN OBERTH DIED IN A HOSPITAL at Nuremburg, West Germany, on 29 December 1989, at 95. He is memorialized by the Hermann Oberth Space Travel Museum in Feucht, Germany, and by the Hermann Oberth Society. There is also a crater on the Moon and asteroid 9253 named after him.

Summary

ALTHOUGH TSIOLKOVSKY, GODDARD AND OBERTH conducted similar research and arrived at similar conclusions, there is no evidence that each knew details of the other's work.

Therefore, NASA considers all three of these scientists as the Fathers of Rocketry.

Konstantin Tsiolkovsky is known for the Rocket Equation, which established the relationship between rocket speed, and the speed of the gas at its exit and the mass of the rocket and its propellants. This equation is the basis of much of the launch vehicle engineering today. Also, in 1929, he published his theory of multistage rockets.

Robert Goddard is known primarily for being the first to develop and fire a liquid propellant rocket in 1926 and later using gyroscopic control and vanes in the engine exhaust for guidance. He also developed pumps suitable for rocket propellants.

Hermann Oberth is known for suggesting regenerative cooling of the combustion chamber and a synergic trajectory, which has a vertical launch to minimize aerodynamic drag, and outside the atmosphere, the trajectory is horizontal to utilize the Earth's rotational speed. He is also known for the Oberth effect, which states the most efficient time for a spacecraft to burn its engines is at the lowest orbital periapsis when its orbital velocity is greatest.

There is one thing they all had in common, however. They were all inspired by Jules Verne's science fiction books, and, above all, they never lost sight of their dreams.

Hermann Julius Oberth
(1894-1989)

Part Four

Sergei Korolev

Chapter 12 - The Gulag

I didn't commit any crime. – Sergei Korolev before being sentenced to the gulag.

Moscow, June 1938

I'LL NEVER FORGET THAT EARLY MORNING of June 27th 1938. Four men, two from the NKVD, the secret police, and two "witnesses," came to my apartment at 28 Konyushkovskaya Street in Moscow. "Are you Sergei Pavlovich Korolev?" they asked.

"Yes, I am."

In just a few minutes, I was taken away. My wife, Xenia Vincentini, didn't have time to give me some clean clothes, nor did the men allow me to say goodbye to my three-year old daughter, Natasha, who was still asleep.

I was taken to Lefortovo prison, where I was interrogated and beaten. I asked for a glass of water, and one of them gave it to me, but then hit me in the head with the water jug. I had recently been given a high position at the Reaction Propulsion Institute, or RNII,[1] and was accused of spending too much money. One of the NKVD men told me, "You're an enemy of the people."

I learned that Ivan Kleimenov, Gyorgi Langemak and Valentin Glushko had been arrested earlier on charges they had worked with anti-Soviet groups in Germany. Their testimony led to my arrest.
Kleimenov wanted to work on solid propellant rockets, and I

wanted to develop a liquid propellant rocket plane. Evidently, this disagreement was behind their testimonies against me. They were arrested along with most of the command staff from German foreign intelligence. Many of them were shot. Both Kleimenov and Langemak were executed, and Glushko was sentenced to eight years in prison. This was all part of Stalin's purges of the scientists, engineers and military leaders. On hearing of their executions, I knew I was in trouble.

One of the guards said to me, "Today is your trial." He and some other guards led me down a hallway to a room. Once inside, the door opened, and the interrogators entered.

"Give me a list of accusations," Kliment Voroshilov, one of Stalin's closest associates, ordered.

"I don't have it," I said.

"Look at your hand!" To my surprise, a piece of paper was in my hand.

I thought surely, Voroshilov would straighten out the matter, but instead, he said, "Do you agree with the accusations?"

"I didn't commit any crime," was my reply. I knew if I confessed, I probably would be executed, which was standard procedure at that time.

Voroshilov shouted, "None of you swine have committed a crime. Ten years hard labor. Go! Next!"

I was sent to several prisons, before I was shipped to the dreaded gulag at Kolyma in eastern Siberia. Millions of people were arrested and sent to these forced labor camps throughout the Soviet Union, and hundreds of thousands died in the brutal conditions in them. Kolyma was one of the worst with a thirty percent death rate.

Now, as I lay close to death, I am free to provide an account of my story to my two favorite cosmonauts, Yuri Gagrin and Alexi

Leonov, for them to make it known to the world.

<><><>

I WAS BORN ON 12 JANUARY 1907[2] in the town of Zhitomir, near Kiev, in the Russian Empire.[3] My father, Pavel Yakovlevich Korolev, was a teacher of literature at a local school. At three years of age, my family fell apart. My mother, Maria Niklovaevna Moskalenko, wanted to attend a school for women, like her sister did, but Father disagreed. As a result, Mother went to live with her sister, and I was sent to live with my grandparents in Nezhin, a town about 160 kilometers to the northeast. Father didn't make much money, so financial difficulties were probably another factor in their breakup. However, I didn't know of the separation until years later. Mother simply told me Father had died.

When I was six, my grandparents took me to an airshow at the local fairgrounds. Sergei Utochkin, who was famous all over Russia for his barnstorming, was there. His flight impressed me, such that, it was a factor for me becoming a pilot later in life.

I didn't have a normal childhood, though. Few children lived near me, and I was very lonely. I took to reading and became a good student in arithmetic and writing. In 1914, life became more difficult for everyone. The tsar's army was mobilized to fight Germany in the Great War, and there was great social unrest. Workers went on strike. They were quickly conscripted into the army, and German prisoners of war took their places.

Mother met Grigory Mikhailovich Balanin while studying at Kiev. He was an electrical engineer who had received his degree in Germany but needed a Russian engineering degree to work here. He was enrolled at Kiev Polytechnic Institute when he met Mother. He married her in November 1916, and they got me from my grandparents. He went to work at the Southwestern Railway in Odessa, on the Black Sea, causing us to move there. He soon became head of the power station, an important position.

Life in Odessa was turbulent. Various groups were fighting for control with the Bolsheviks prevailing by early 1920. After things became normal, I was enrolled in Professional Building School Number 1, where I was trained to be a roofer and tile layer. The school was quite unusual. Its faculty included physicists and a mathematician, so I was trained in strength of materials, physics and mathematics in addition to a manual vocation.

I was fascinated by the Odessa hydroplane squadron stationed nearby. I would often swim to the base, hold onto the barbed wire and watch the planes. One day, while I was watching, a mechanic yelled to me, "Well, what are you hanging around for? Why don't you give me a hand? Can't you see I'm having trouble with this motor?"

From then on, I was often at the base. The eight patched-up M-9 biplanes became my passion. Finally, the squadron commander, Alexander Shlyapnikov, gave me a ride. We flew out over the breakwater, climbing into the wind. There were tiny people and toy ships below. A thrill I'll never forget. I didn't tell Mother, but one day when she commented about the beauty of the clouds, I said, "Oh Mother, if you could only see them from the top."

Mother didn't stop me from flying as a passenger. However, in the summer of 1923, I had a thrill far more than I wanted. I was flying as a passenger, with Brzhezovsky as the pilot of the seaplane, when the engine cut out. I went out on the wing to check the oil supply, but suddenly the plane shook violently, and I fell into the sea. Fortunately, I only dropped about ten meters and wasn't hurt. Brzhezovsky landed near me and picked me up.

My friend and schoolmate, N.Y. Abezguz, once said to me, "You seem to have developed a clear goal in life – aviation."

I replied, "Anything that doesn't relate to my work in the port's glider club and the province's aviation sport section or to my theoretical studies is a distraction to me."

I had a good relationship with my stepfather as he always encouraged me in my studies. However, my interest in aviation caused me to let some of the academics slip. My mathematics teacher complained about my lack of concentration. This caused my stepfather to lecture me, "Education and a profession will make you independent. Remember that. Handstands, yachts and aeroplanes are rubbish."

His remarks didn't deter me. I was serious about aviation. I also had learned how to use a slide rule and drafting instruments, which I used to design a glider. I named it the K-5. It was accepted for construction by the OAVUK[4] in 1924, when I was only seventeen.

It was at school that I met Xenia Vincentini. She was beautiful, slender with a braid of hair that went below her waist and had big brown eyes. We both graduated in 1924. I asked her to marry me, but she refused. Instead, she wanted to study at the Chemical and Pharmaceutical Institute in Odessa. I applied to the Zhukovsky Academy in Moscow, where aeronautical greats of the time were teaching. Andrei Tupolev was one of them. He taught while designing military aircraft for the government.

There was a problem with my application to Zhukovsky. I didn't meet the requirement of being a certified military pilot, so I had asked for a waiver on that prerequisite Xenia had turned me down, and now the academy delayed my application. I was very discouraged. However, I heard Kiev Polytechnic Institute had started an aviation section, so I enrolled there.

At Kiev Poly, I took courses in math, physics, chemistry, mechanics, strength of materials, thermodynamics, machine parts, electricity, architecture and building, statics, and political economy. In addition, I took a lubrication workshop, among many workshops, and I worked as a fireman on a locomotive. A course on glider construction was added in 1925. As a first-year student, I wasn't selected to work on gliders, but I was able to fly a finished aeroplane. Thus, I became a pilot, before graduating in June 1926.

I went on to the Moscow Higher Technical University.[5] There, I became involved in a local amateur gliding society. We even flew during the cold of winter and into the spring of 1927. In the fall of that year, I finally got to fly in the famous Koktebel competition. Flying a glider in that event was a heady experience. As a result, Savva Lyushin, a classmate, and I began designing our own glider for the competition. I told him, “It must be reliable, even if we had to settle for a less than optimum aerodynamics.”

We worked on the *Koktebel*, naming it after the event itself, through the summer of 1929. I did take time out in August to solo in a powered biplane, an Avro 504K, then fly as much as possible the rest of the month.

In October, I flew the glider in the Koktebel competition in the Crimea. I wrote to Mother, “I felt a colossal sense of satisfaction and wanted to shout something into the wind that kissed my face and made the bird tremble with its gusts. It’s hard to believe that such a heavy piece of metal can fly. But it’s enough to leave the ground and feel how the machine became alive and fly whistling, answering to the least movement of the controls. It’s the greatest satisfaction and reward to fly by oneself in one’s own machine.”

The journal, A*eroplane*, gave us considerable praise: “The intention of the designers, Lyushin and Korolev, was to build a machine very stable in the longitudinal direction that would not fatigue the pilot in long flights. In this they have been fully successful.”

Earlier that year, the Communist Party recommended engineers’ education be accelerated in order to meet the country’s needs. Moscow Higher Technical University responded to this by reducing the number of exams needed for a diploma if a practical design were substituted. As a result, I proposed a glider design I had partially worked previously. This was accepted and Andrei Tupolev became my advisor on the design.

“You are one of the easiest students I have,” Tupolev told me. “I just need to review your work and make some minor suggestions here and there.”

I defended my design in December 1929 and a few weeks later

received my certification as a qualified aero-mechanical engineer. I went to work at Factory 22, in Fili, a village just outside of Moscow, where I worked on aeronautical components, but I was more interested in building gliders and powered aeroplanes, which I did in my spare time.

About that time, I was working at the Central Aerohydrodynamic Institute, TsAGI, and met Friedrikh Tsander, an amazing engineer. At TsAGI, I was the lead engineer in the development of an autopilot for Tupolev's heavy bomber, the TB-3. Both of us became passionate about the newly formed rocket group in Moscow, the Group for Studying Reaction Propulsion, known as GIRD.

GIRD was developing a rather crude rocket engine, the OR-1. Some called it a "reconstructed soldering lamp" as it burned gasoline and air and only generated about five kilograms of thrust. After this, Tsander, with a friend at TsAGI called Yuri Pobedonostsev, and I developed a much more sophisticated engine, the OR-2. The OR-2 used gasoline and liquid oxygen as propellants. It had a combustion chamber cooled by liquid oxygen, and the nozzle was cooled by circulating water. It developed 50 kilograms of thrust.

Things were very busy at GIRD, but all along I still found time to write and visit Xenia. Finally, on 6 August 1931, we were married in Moscow with only two friends present. After the wedding, she immediately returned to Kharkov, where she worked as a district health inspector, to arrange her transfer to Moscow.

On 17 August 1933, we launched the GIRD-09, which was designed by Mikhail Tikhonravov. It was the first successful launch of a liquid propellant rocket in the Soviet Union. GIRD-09 reached an altitude of 400 meters. I put a poster on the wall stating the importance of the launch, and I ended the text with, "Soviet rockets must conquer space!"

Mikhail Tukhachevsky, a Russian military man, had been funding the Gas Dynamics Laboratory, GDL, but on hearing of GIRD's successful launch began funding us as well. He saw the potential of rockets for military purposes and was instrumental, in late 1933, in combining GIRD and GDL into one organization, the Reaction Propulsion Institute, or RNII. The military engineer Ivan Kleimenov headed up RNII, and I became the Deputy Chief.

In 1934, we began work on winged rockets for weapon systems, as they had about five times the range of a ballistic missile.[6] However, the main technical problem we encountered was guidance and control. I told one young engineer who came to work for me, "Well, all you young people necessarily want to build rockets or rocket motors, and believe they constitute the full matter. We already know how to build rockets and motors, but guidance of the flight and stability of movement have become a bottleneck."

I headed up the team that developed the GPS-2 and GPS-3 automatic gyroscopic stabilizers with two and three degrees of freedom.[7] These devices had to be adjusted to the aerodynamic characteristics of the rocket and its intended trajectory. It was demanding work, and we sometimes worked twenty-four hours straight.

We developed a series of rocket-propelled gliders through 1938 culminating in a glider powered by Glushko's ORM-65 rocket engine. The ORM-65 used nitric acid and kerosene for propellants and produced 175 kilograms of thrust.

All of this seemed to be for naught after my arrest, but, in fact, it was what saved me from dying in the gulag.

Sergei Korolev in cockpit of glider. (Photo from Star Child site of Astrophysics Science Division (ASD) at NASA Goddard Space Flight Center)

Chapter 13 - The V-2 in Russia

There is no such thing as an unsolvable problem. – Sergei Korolev

Kolyma Gulag, September 1939

THE KOLYMA GULAG HAD A SURFACE GOLD MINE where there was a forced labor camp to remove the gold. I was given the task of cutting trees, digging, and pushing a wheelbarrow, all part of the process to extract the precious metal for the Soviet Union's needs. It was hard work. I felt like a beast of burden more than a man. Besides the harsh labor, most of us prisoners received little food, and our clothing was falling apart. We were crowded into houses that retained little of the scant amount of heat the stoves provided, and sanitation was non-existent. Many died of overwork, starvation, disease, and brutal treatment. Some were executed for petty reasons.

I longed to be with Xenia, but I feared I would never see her and Natasha again. I thought, before the winter is over, I'll be dead, frequently came to mind. When Natasha was born, I just knew she would be the girl I wanted. I named her after Natasha Rostov in the novel *War and Peace*. I wished I had spent more time with her, but Xenia and I had our professions to tend to. At that time, we couldn't get an apartment of our own, so we lived with my mother and stepfather in Moscow. They helped raise Natasha. But that was the past. I just needed to survive the winter and the ill treatment.

<><><>

SURPRISINGLY, IN SEPTEMBER 1940, a guard came to me

and said, "You are being transferred to another prison."

I was to go to Magdala, about 150 kilometers away to board a ship to a prison in Moscow. My fellow prisoners collected clothing for me, and I began walking toward the town. I stopped a truck and asked for a ride. The driver agreed only if I gave him my sweater. In Siberia, there was nowhere else to go, so I resigned myself to going to another prison.

I spent the winter at Magdala, as I just missed the ship out of the Okhotsk Sea.[1] Later I learned everyone on the ship was lost at sea in a severe storm. Winter came soon and was terrible for me. The temperature typically was about minus 45 degrees C., and I had to find shelter. I was kicked out of one army barracks after another, until I learned to hide under the cots. I worked as a laborer, shoe repairman and did any odd job I could find. Bread was provided for us outside the barracks, but no one knew where it came from.

In the spring, I left Magdala on a train, but on the way to Moscow, I contracted scurvy. My body was swollen, and I had lost most of my teeth. I was thrown off the train because of my condition.

I was almost dead when a kind old man found me and said, "I'll make you better, sonny."

He fed me and tended to my gums to stop the bleeding. After about a week or so, I was on the train again headed for Moscow.

Later, I learned Stalin recognized the importance of aeronautical engineers due to the impending war with Hitler. He set up a system of prison design bureaus, called sharashkas, to exploit prisoners with technical ability.

Sergei Tupolev's request for my services in his design bureau in Moscow, the KB-29 *sharashka*, saved me from almost certain death in the gulag. Ironically, Tupolev was also a prisoner. At this time, I worked mainly on aircraft he designed.

In 1942, because of the German invasion of Russia, Tupolev's team was evacuated to Omsk, in Siberia.[2] I went with them, and did some rocket designs for an aerial torpedo. However, in November, I was transferred to a *sharashka* in Kazan.[3] There, I worked with several aircraft jet engine experts, including my former accuser, Valentin Glushko. In January1943, I became chief designer for the development of his jet engines, and I was given the opportunity to develop a rocket engine for the Pe-2 dive-bomber.

We worked on this engine, the RD-1, through October 1943, when it was flight-tested. The engine operated for two minutes and increased the aircraft's speed from 340 to 420 kilometers per hour.[4]

In the summer of 1944, the entire experimental design was released from custody, and I was discharged with all my convictions removed from the records. Part of my sentencing, after my arrest, was the right of communication while in prison. After my release from the gulag, I began sending letters to Xenia. Now, I could actually see her when I returned to Moscow.

At this time, the design bureau was put under the People's Commissariat of Aviation, which was headed by the NKVD. Glushko was made chief designer, and I was his deputy. During the ensuing year, we remained at Kazan and developed the D-1 and D-2 rockets. The D-1 had a launch weight of 1,100 kilograms, including a 30-kilogram warhead, and had a range of 12 to 30 kilometers. The D-2 weighed 100 kilograms more, with the same warhead but had a range up to 70 kilometers.

In July 1944, Winston Churchill informed Josef Stalin about the V-2 rocket the Germans had developed. As a result, British and Soviet engineers met in Poland to review the pieces of several V-2s. Some parts were sent back to Moscow, where one of the engineers, on seeing the V-2 components, made the comment, "This is what cannot be."

One of our engine experts said, "This is a profound discovery. The V-2 engine developed twenty tons of thrust, whereas our most powerful engines could only approach producing one and one half tons. The V-2 had a range of 300 kilometers, more than four times anything we developed. The Germans evidently put a lot of time and money into its development, which is very surprising with a war going on."

Another surprise for us was the propellants the V-2 engines used. Instead of using nitric acid and kerosene, as we usually did, the Germans used liquid oxygen and alcohol.

In late summer 1945, I was commissioned a colonel in the army, and I flew to Germany with the team to gather more information on the V-2. I wore my new army uniform and drove a German Opel to the office of Boris Chertok, who headed the Rabe Institute there.

"I'm told you want to see everything we're doing regarding the German rocket," Boris said.

Instead of answering, I noticed his secretary and asked, "Why does a Soviet official have such a pretty German secretary?"

"She can type and do stenography, and she speaks both German and Russian."

"She is pretty," I observed once again.

"Ironically, we're allowed to use Germans, even if they've been Nazis, but we can't use Russian prisoners of war." Boris explained, then changed the topic, "At the Institute, we solve problems every day. We first understand the functioning of each V-2 component, then reconstruct the documentation and make drawings of it. Finally, we build the device. As you probably know, we're reconstructing other parts of the V-2 elsewhere in Germany. Would you like to visit our Institute and see what we've done here?"

"No, thank you. I'm in a hurry to go to Berlin."

In Berlin, I learned that the British had invited American and

Russian groups to observe V-2 launches at Cuxhaven, on the North Sea near the mouth of the Elbe River. This was part of Operation Backfire, an operation set up by the Allies to evaluate German technology. I wasn't invited, but I went anyway. I was stopped at the gate and forced to stay outside the fence during the whole demonstration, while Glushko and two other Russians were part of the official delegation inside. I was beside myself that they hadn't invited me to be there.

Knowing the value of the V-2, I interviewed dozens of engineers and technicians to get a complete understanding of the rocket.

In May 1946, I had Xenia and Natasha join me in Germany. We had a marvelous time together through much of the summer. We visited many interesting places and even took a trip around the country in the Opel.

Stalin was annoyed that most of the leaders of the German rocket development team, including von Braun, had gone over to the Americans with many V-2s and their components going to the Americans as well. On 13 May 1946, Stalin signed the decree beginning development of Soviet ballistic missiles. The Minister of Armaments, Dmitiri Ustinov, was put in charge. In August 1946, the Scientific Research Institute, NII-88, was set up, and I was made chief designer of SKB-1 and was responsible for the development of long-range ballistic missiles. Valentin Glushko was named chief designer of OKB-456, the rocket engine plant at Khimky, near Moscow.

Next, more than 200 German employees captured at the Mittelwerke V-2 factory in Germany were rounded up on the night of 22-23 October 1946 and sent by rail to Gordodomliya on Lake Seleger, between Moscow and Leningrad. One of the German leaders, Kurt Magnus, complained to me, "I'm about to get married, but I don't have the documents to marry yet."

"You can marry without any problem in the Soviet Union," I said.

"You can go back to Bleicherode in a few weeks."

"But do you really believe that we can travel back to Bleicherode? After what we've experienced today, I doubt it."

"I give you my word of honor as a Soviet officer."

Magnus said nothing more and went with the others. I never heard whether he went back to Germany.

There was another incident, which occurred before the mass movement, that was quite humorous. Helmut Gröttrup's wife refused to go because she heard there was a famine in Russia, although she was promised the use of a farm there. She finally agreed to go when she was told her cow could go, too. Our men put her cow and hay into a railroad car, along with the belongings of the other families. The Germans were treated quite well at Gordodomliya.

In Russia, the Germans taught us everything about the manufacture and assembly of the V-2. In the summer of 1947, a contingent of Germans and Russians with several V-2s were sent to Kapustin Yar, in southern Russia. On 18 October 1947, a V-2 was launched, and two days later, a second one was launched. They went way off course. The men joked, "It went towards the city of Saratov."

Ustinov, who was at the launches, didn't see the humor. He told Dr. Magnus and Dr. Hoch, experts in gyroscopes, guidance and control, "This is your missile, figure out what went wrong."

The problem was traced to gyroscope error caused by vibration. The problem was quickly resolved, and the next flight's trajectory was near perfect.

Ustinov directed, "All German specialists and their assistants be given 15,000 rubles each and a canister of alcohol."

Needless to say, there was a big celebration.

However, the launches were with German V-2s, assembled by

Soviets. The next step was to launch a Soviet manufactured V-2, the R-1. The R-1 had a RD-100 engine, essentially a V-2 engine made with Russian materials. The first launch, on 7 September 1948, had a control system failure, however a month later, on October 10th, we had a successful launch. There were technical problems, however by 1950, the R-1 was commissioned for service.

In mid-1947, it was obvious that something better than the V-2 was needed. A rocket with a three-hundred-kilometer range was not enough. Twice the range was the new design requirement. I proposed the R-2, and the German leader, Gröttrup, proposed the G-1. We would be in fierce competition until one design was finally selected.

Sergei Pavlovich Korolev (1907-1966)
(Photo from the Public Domain)

Chapter 14 - The Road to Sputnik

The Soviet Union has become the seacoast of the universe.
– Sergei Korolev

Kapustin Yar, 1948

AFTER WORLD WAR II, there was much work needed to develop the capability to launch an intercontinental ballistic missile, particularly one that could put a satellite in orbit. Much of this activity was for military purposes, but it also served to let the world know the Soviet space program was very advanced. Each step on the road to *Sputnik* was a giant leap in space technology. We started with the V-2, which had a range of about 300 kilometers, and manufactured our version of it, the R-1.

The R-1A was the first step beyond the R-1. It was simply an R-1 modified to separate the payload in flight. Two launches occurred in May 1948 at Kapustin Yar, which was on the Volga River just east of Stalingrad[1]. Initially, the payload was damaged from the shock of the parachute opening. However, on the second launch, the parachute was gentler on the payload, and we were able to recover it undamaged.

I reported, "From these flights, we were able to experimentally show that it was possible to transport equipment for investigation of the upper layers of the atmosphere by rocket to altitudes of 100 kilometers, to eject the equipment container, and to safely bring it back to Earth."
At the same time the R-1A was being launched, we launched two more series of R-1s at Kapustin Yar. They had varying degrees of

success, but served to instill confidence in the Soviet leaders regarding the new rocket industry. These flights also were the basis of developing an operational infrastructure for launching rockets as weapons of war. In November 1950, the R-1 was formally adopted as armament by our armed forces.

It was during these early years that I gave lectures to students at my alma mater, MVTU, and I shared some sensitive, highly classified information with the students. The lectures also included some of my ideas for a long-range rocket. At that time, I introduced the practice of hiring students while they were still at the university to do their diploma work and eventually sifting out the best of them for permanent positions after graduation.

One time I told a student, "I won't take you alone, but I'll take you and some good guys you select from your class." Fourteen of them came to work for me.

In 1946, I saw my daughter reading Jules Verne, as I had done so many years before. I told her, "You know, in about thirty years a human will be on the Moon."

"Such a feat is impossible," she argued.

"Please remember this day. I tell you, man will walk on the Moon," was my response.

<><><>

THE RANGE OF THE R-1 WAS INSUFFICIENT for Stalin and the Soviet military. In May 1947, the German team, under Helmut Gröttrup, was given the task of designing the G-1 with twice the range, 600 kilometers. The G-1 was a single-stage design with radio-controlled cut-off for improved accuracy and a separable warhead. Duralumin alloy[2] construction throughout the airframe and the integrated propellant tanks[3] reduced its empty weight by forty percent relative to the V-2.

At the same time, I designed the R-2.

In December 1948, the Germans defended their design before NII-88, the Scientific Research Institute. The government favored the G-1 over my R-2, but I argued, “We cannot immediately master the manufacturing technology of the G-1.” Thus, I convinced them to develop the R-2 for production. However, I was forced to use some of the German concepts, namely a fuel tank that served as part of the vehicle structure, their cut-off system and a separable warhead.

On 25 September 1949, the first of five R-2 launches were made at Kapustin Yar. The deployment of the warhead was marginally successful due to a malfunction of the automatic stabilization system at the time of separation. This problem was corrected in later launches. The last flight test was in July 1951.

The R-2 had about twice the range of the R-1 and was equipped with a radiological warhead.[4] The engine[5] was developed without German involvement. It had thirty-six percent more thrust than the R-1 engine, mainly due to a higher chamber pressure and a fuel with a greater concentration of alcohol.[6] Like its predecessors, it used hydrogen peroxide which when combined with a catalyst provided steam to the turbine that drove the propellant pumps. I told the designers, “Use methyl alcohol,[7] for fuel for the main engine, not ethyl alcohol. The launch troops are drinking up all the rocket fuel!”

<><><>

WHILE I WAS DEEPLY INVOLVED in the R-1 and R-2 projects, Xenia was very engrossed in her profession. As a result, we saw little of each other, and we drifted apart. I had met an English translator, who also lived in my apartment complex. Nina was young and very attractive. She listened intently as I described some of my problems to her. We suited each other and immediately fell in love.

Xenia was furious when she learned of Nina. She screamed, “You miserable husband. How could you do this?” But, it was too late.

Xenia and I divorced in 1948, and I married Nina Ivanovna Kotenkova the following year. As a result of the divorce and my work, I saw little of Natasha.

I didn't drink much alcohol, and I chose to live a rather austere life. However, I was fond of women, and they seemed to like me. This did not please Nina, yet she lived with my indiscretions.

The next step in Soviet rocketry was a leap to an intermediate range missile, one with 3,000 kilometers range. This was to satisfy Stalin's directive to have a missile capable of reaching anywhere in Western Europe. Again, the Germans, with their G-4, and I, with my R-3, were in competition to create a rocket for these requirements. Glushko designed the same engine for the G-4 and the R-3.[8]

For this new missile, we studied a winged cruise missile and multi-staged ballistic missiles. The cruise missile concept had more range, but it operated entirely in the atmosphere, which made it vulnerable to enemy anti-missile systems. Thus, we abandoned it.

Glushko built his engine, which used liquid oxygen and kerosene propellants for higher performance. The high-pressure chamber, fed by nineteen preburners, presented manufacturing problems and had unresolvable synchronization problems. Therefore, this engine never was used, but it served as the basis of future engine designs.

The NII-88 met in December 1949 and subjected my R-3 design to intense criticism. They preferred Gröttrup's G-4 design. After I had some heated discussions with NII-88, they approved further development of technology for the R-3, but not the missile itself due to it being such a leap in technology.

Instead, they decided on evolving these technologies in a smaller R-3A, starting with a more conservative liquid oxygen and alcohol engine.[9] This became a long series of design studies with no vehicle ever built, but the R-3A became the basis of the R-5M.

The German team was unaware of our work, and they were

gradually phased out. Members of their team returned to Germany beginning in April 1951. By October 1951, their work was stopped, and by November 1953, the last member of the group returned to his homeland. The German designs and design concepts were a major part of our vehicles through the R-7 and beyond. However, Glushko departed early from the German engine design and went forward on his own.

Glushko and I had many disagreements, particularly on the type of propellants. I favored liquid oxygen and kerosene, but he, like the military, wanted to use storable, highly toxic propellants, such as nitric acid and hydrazine.

Following years of design studies, approval came in February 1953 for development of an intercontinental ballistic missile with a range of 7,000 kilometers.

In order to concentrate on development of an intercontinental missile, I spun off intermediate range missile development to a new design bureau at Dnepropetrovsk in Ukraine headed by my assistant, Mikhail Yangel. Thus, the R-5M[10] was developed and produced there. It had a range of 1,200 kilometers, and it was the first missile with an atomic bomb in its warhead. Liquid oxygen and alcohol were the propellants used by the engine.[11]

Gröttrup's concept, a cluster of five G-4s to achieve intercontinental range, was the basis of the R-7[12]. It had four booster engines and a core sustainer engine to take over when the boosters dropped off. However, I conceived a core sustainer stage, which would allow all engines to be ignited on the launch pad, eliminating the problems of an air start of the core engine in Gröttrup's design.

The R-7 had the liquid oxygen tanks in the forward section to eliminate precooling the instrumentation. The fins were eliminated to reduce aerodynamic drag. Small steering rockets replaced the fins for directional control. Both propellant tanks formed the structure and the skin of the rocket. And an aluminum-magnesium alloy was used to reduce the vehicle weight.

Gröttrup had proposed a blunt nosed vehicle to quickly decelerate the vehicle on reentry into the atmosphere, which would reduce aerodynamic heating. However, it was decided that a pointed nose provided a faster reentry into the atmosphere with less chance of being intercepted by an enemy missile. The heating problem on reentry with this design would eventually be overcome.

The R-7 faced enormous development problems due to Glushko's engines. The entire vehicle was quite large to meet the range and payload requirements. As a result, the engines were large, too. Combustion instability increases as the chamber size increases. This is a fundamental law of physics. I told Glushko, "The problem has to be resolved as the resulting vibrations from pulsating combustion will damage or destroy the engine."

Glushko's replaced the troublesome single-chamber engines with a very complex engine design. Each of the engines[13] had four small chamber-nozzles sharing common propellant pumps. This achieved stable combustion. I wasn't satisfied with the complexity of Glushko's design, but it was the only way to eliminate combustion instability.

By attention to every detail, we overcame all obstacles. I personally checked all aspects of the vehicle assembly and launch preparations. We worked twelve- to fourteen-hour days seven days a week. I was always open to new ideas and never interrupted the person offering one. As a result, the final design didn't look much like the original proposal.

<><><>

DURING THE R-7 DEVELOPMENT, I followed the American journals on space technology. There was a large amount of unclassified information about the U.S. Atlas launch vehicle. Although I didn't agree with the design, I used it to motivate the men, "Look, those Americans are going to beat us to orbit," I often said. I noticed the R-7 was much larger than the Atlas. I traced the main difference to our warhead being much heavier.[14]

The thought of the Americans and us having intercontinental ballistic missiles with nuclear warheads made me pessimistic. I often said, "We will all vanish without a trace."

We had five unsuccessful flight tests of the R-7 from May through July of 1957. The selected test site was Baikonur, in the desert of Kazakhstan east of the Aral Sea. We didn't like being so far from home, but there were definite advantages to the site.

The flat terrain facilitated the radio control system of the rocket, which required receiving uninterrupted signals from ground stations hundreds of kilometers away, and the missile trajectory had to be away from populated areas. Also, Baikonur was closer to the equator, where the Earth's surface has a higher rotational speed for additional launch speed.

The conditions at Baikonur were harsh. I was given a simple cabin, but the men lived in tents and were subjected to temperatures from below zero to above 38 degrees C.

The first test was a total failure, due to axial vibration of the vehicle, called "the pogo effect." This phenomenon is named after a child's pogo stick because of the motion of the vehicle when it occurs. The effect occurs when the propellant feed system resonates with the thrust oscillations of the engine. It is an ever-increasing vibration which causes loose connections, failed tubes, etc. Pogo suppressors in the feed lines were eventually added to prevent this problem. There were three launch attempts aborted on the pad due to a faulty valve and we had another total failure just after launch.

On 21 August 1957, we had a successful launch, which achieved a range of 7,000 kilometers. A 1,400-kilogram satellite was to have been launched by the R-7 on 4 October 1957, but it was not ready in time. Thus, we hurriedly developed *Sputnik* as a replacement.

Sputnik[15] was a small sphere, about twice the diameter of a basketball. It weighed 83 kilograms and was made of highly polished aluminum alloy. It had just one watt of power from three silver-zinc batteries, producing a 0.4 second duration radio signal. Four antennas were deployed at 35-degree angles. The satellite was built without drawings due to the quick schedule. I was everywhere, supervising all aspects of its construction.

On October 4th, *Sputnik* became the first artificial satellite of the Earth, and it shocked the world. It circled Earth once every ninety-six minutes, crossing over America seven times a day. One London paper stated, “Russians Win Race to Launch Earth Satellite.” Many Americans feared more sinister uses of our new space technology, which apparently was ahead of theirs. They were very fearful of a Soviet military presence in space.

Sputnik functioned for 21 days. I was elated over its success and used the victory to promote my next major advance. My proposal resulted in bitter disputes, until a compromise was reached.

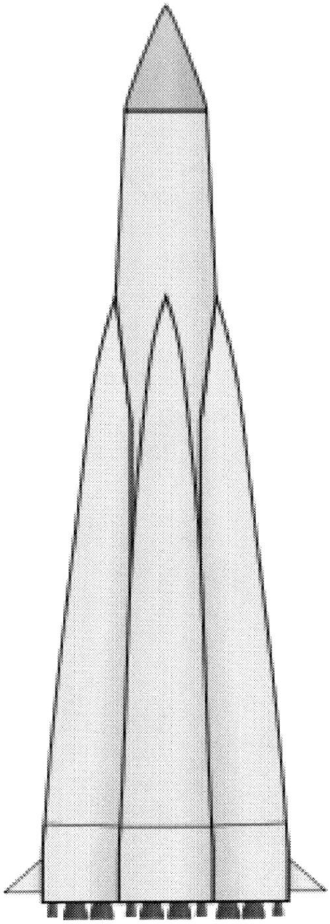

Sketch of R-7 for Sputnik I launch.
(Provided by GW_Simulations)

Chapter 15 - Luna and Vostok

To be the first to enter the cosmos, to engage, single-handed, in an unprecedented duel with nature – could one dream of anything more? – Yuri Gagarin

Kaliningrad, 1960

I FIRST MET CHIEF DESIGNER KOROLEV on 18 June 1960 when he summoned twenty of us to his design bureau in Kaliningrad.[1] He showed us an actual spaceship and said, "What we're doing is really the easiest thing in the world. We invent something, find the right people to build it properly, and place lots of orders for components with the best and most experienced factories around the country. When at last they've delivered what we've ordered, all we have to do is put the pieces together. It's not very complicated."

We, cosmonaut candidates, knew there was much more to building a spaceship, but we were touched by his warmth and friendliness. I listened intently to everything he had to say and had many questions on space and rockets. He asked me, "What is your name?"

"Yuri Gagarin."

"Yuri, stand up. Tell me, my little eagle, about your life and family."
I spent the next fifteen minutes telling the Chief Designer about myself. I told him I grew up on a farm and went to a technical

school before being drafted into the Soviet Army and being picked for flight school and becoming a MiG-15 pilot.

He listened intently and forgot about the rest of the cosmonauts. I could tell it was the beginning of a lasting relationship.

The Chief Designer was anonymous to all but the insiders in our country and unknown to everyone in the Western world. His name was kept secret, due to fear of his assignation by Western agents. He was addressed within the Soviet space community as “The King,” “The Boss of Bosses,” or affectionately, by his first two initials, “SP.” The press and radio announcers spoke of him as “The Chief Designer.” After I got to know him, I called him SP.

<><><>

THE FACT THAT WE WERE ALL GATHERED at his design bureau was an achievement in itself as the Chief Designer had considerable opposition after *Sputnik 1*. At first, the Earth satellite made little news in our country. Initially, Secretary Nikita Khrushchev[2] hardly paid attention to the accomplishment. However, the reaction of the Americans and others brought to our leaders’ attention the propaganda value of *Sputnik*.

Seeing his opportunity, the Chief Designer followed *Sputnik* with a proposal to put a man in space with top priority. He encountered fierce opposition from the military and others. They wanted further rocket development to be directed toward ballistic missiles with nuclear warheads and for satellite reconnaissance. Finally, a compromise was reached where a man in space program would be combined with rocket development for military purposes.

Khrushchev told the Chief Designer, “Do something bold for the upcoming fortieth anniversary of the Revolution next month.”

Sputnik 2 went up on 3 November 1957. In its 500-kilogram payload it carried the dog Laika[3], and Khrushchev boasted of the Soviet lead in space at the anniversary celebration.

The Americans once again were shocked and could see where the Soviet space program was heading. The U.S. launched *Vanguard* in December, but it failed shortly after liftoff. However, on 31 January 1958, they successfully launched *Explorer 1*. It discovered and recorded a radiation belt above the Earth. It was named the Van Allen Belt, after James Van Allen who confirmed its existence by analyzing the data from *Explorer 1*.

A few days later, we tried to launch another *Sputnik*, but it failed. It wasn't until May that *Sputnik 3* went up. It weighed 1,327 kilograms, as it carried a large amount of scientific equipment, but it didn't map the Van Allen belt due to a faulty tape recorder.

Everyone knew it was just a matter of time before an attempt was made to send an unmanned vehicle to the Moon. Several had proposed this mission even before *Sputnik 1*. In 1955, SP had proposed a three-stage R-7 to reach the lunar surface. After the *Sputnik* successes, SP wrote a paper proposing robotic missions to the Moon. "First studies of the Moon and interplanetary space, at distances from Earth of 400-500 thousand kilometers, will also create the necessary prerequisites and premises for man's penetration into interplanetary space, the Moon and the planets."

Thus, during 1958, there were a total of seven attempts by both the Americans and us to send an unmanned probe to the Moon. None were successful.[4]

Finally, in January 1959, SP's team sent *Luna 1* up, using a three-stage version of the R-7. The upper stage engine was designed by SP's design bureau with another design bureau. It missed the Moon by about 6,000 kilometers,[5] but it determined the Moon had no magnetic field. It was the first man-made object to go into solar orbit.

Luna 1 was publicized as a planned flyby of the Moon, when, in fact, it missed impacting the Moon due to a third stage separation error. In September, *Luna 2* impacted the Moon, another first by the Soviet Union. Three weeks later, on the first anniversary of *Sputnik 1*, *Luna 3* photographed the far side of the Moon, yet another first by the USSR.

During these launches, SP's personality became more apparent. He was a very complex person. He understood that designers make mistakes. As long as the designer acknowledged his fallibility, SP understood. He didn't like people who tried to cover up their mistakes, even if they were small errors. He punished them by swearing and shouting, which wasn't always justified. The next day, he would be embarrassed if his tirade was wrong, not to the extent he would apologize, but he would be extra warm and friendly. In a fit of anger, he would fire someone on the spot, only to rehire the person the next day. He was very strong willed and very decisive.

He did have several teams working on the same problem to be as sure of the solution as possible before making a decision.

In November 1958, the Council of Chief Designers approved the *Vostok* manned space program, in combination with the *Zenit* spy satellite program. The Chief Designer was to head up *Vostok*.

<><><>

ON THAT DAY IN JUNE 1960, The King led us into the heart of his design bureau, OKB-1. He showed us a room with a dozen spacecraft in various stages of construction. As pilots, we were overwhelmed by the strangeness of the vehicles. We couldn't grasp the function of the equipment. There were no wings, no control surfaces, no propulsion system and no obvious landing gear. They were spheres, which he explained were *Vostok* space capsules for us to ride in. SP had each of us get in a spacecraft before we left his design bureau.

<><><>

IN OCTOBER 1960, SP had two Mars probes fail. Khrushchev, wanting to impress the world with a successful Mars flyby, turned to Mikhail Yangel to use the R-16 he had designed for the Mars shot. The R-16 used nitric acid and hydrazine as propellants. They had the advantage of being storable in the rocket until launch, whereas the liquid oxygen used in the R-7 took several hours to load and if the launch was delayed the oxygen boiled off and had to be replenished. However, the R-16 propellants were very corrosive and highly toxic.

Khrushchev sent Marshall Mitrifan Nedelin, Commander in Chief of the Strategic Missile Force, to supervise the R-16 launch. On 23 October, as the zero-hour approached, nitric acid began leaking from the vehicle. Nedelin, who had taken over supervising the launch, instead of draining the propellants and purging the system with nitrogen, sent dozens of ground personnel to eliminate the leaks. He also refused to cancel the firing sequence system, instead he had it revised and delayed.

A wrong electrical signal went to the upper stage causing it to fire. The flames from the upper stage caused the lower stage to explode killing the ground personnel near it. The upper stage lost its support and crashed to the ground spilling fuel everywhere, which quickly caught fire. More ground personnel were incinerated. More than 190 personnel lost their lives, including Nedelin.

The public was told it was an "aircraft accident." However, we cosmonauts were told that a prototype missile had blown up, and it was not one of SP's R-7s. The sudden disappearance of so many people we knew raised many questions, and, eventually, we learned the truth.

Tragedy hit us once again just three weeks before the planned manned flight into space. Our youngest cosmonaut, Valentin Bondarenko, was undergoing testing for the effects of being in space for a long period. He was in an isolation chamber for fifteen days, but as he was coming out of the oxygen-rich chamber,[6] a fire started, cause unknown. When he was removed, there were burns over his entire body. He died a few hours later.

SP was worried about weightlessness. He had each of us sit behind the pilot of a MiG-15. The pilot put the plane through a parabolic arc, which simulated weightlessness for about thirty seconds. It gave us some idea of the effect but was nothing like the real thing.

He was also concerned the retro-rockets would not fire and would brake the spacecraft thus leaving us stranded in space. For this reason, he had ten days of oxygen, food and water on-board the spacecraft, which would allow time to resolve the problem, hopefully. On the other hand, he was worried where we would land and could we survive the landing. To set an international altitude record, a cosmonaut had to remain in the vehicle through the landing. To ensure the cosmonaut would survive without injuries, an ejection system with a parachute system was devised for a safer touchdown.

Due to the higher risk of a complete orbit, compared to a sub-orbital flight to a very high altitude, SP's designers didn't want the first manned spacecraft to circle the Earth. SP, however, wanted to beat the Americans by a large amount, so he decided on an orbital flight.

We trained long and hard for the mission. We went through the scheduled sequence of events for the flight and several emergency procedures including use of the ejection system. We repeated everything until it became second nature to us. In the end, the cosmonaut to go on the first orbital flight became a competition between Gherman Titov and me.

Chief of Cosmonaut Training Nikolai Kamanin watched us carefully as he would make the final selection. He said, "Titov is of a stronger character. The only thing that keeps me away from deciding in his favor is the necessity to keep a stronger cosmonaut for a 24-hour flight. It's hard to decide which of them should be sent to die, and it's equally hard to decide which of these decent men should be made famous worldwide."

The R-7 was moved to the launch pad at 5:00 AM on the morning of 11 April 1961. SP followed it all the way to the pad. At one in the afternoon, SP escorted Titov, my backup, and me to the top of the gantry. We rehearsed the boarding procedures one more time.

That evening, Titov and I went to our cottage, and Kamanin followed us. We were all nervous about the flight the next day. SP joined us, and seeing us so tense, teased, "I don't know what all this fuss is about. Five years from now, the unions will be subsidizing holidays in space."

We all laughed. SP looked at his watch and left, which meant we should bed down.

None of us slept well that night. In the morning of 12 April, we had breakfast, were examined by the doctors and donned our spacesuits. Under our suits, we were wired to monitor our physical condition during the flight. People asked for my autograph. I couldn't refuse them, so I spent several minutes signing papers handed to me. Then, I climbed into the *Vostok*.

There was a slight delay in the countdown due to a minor electrical problem. SP came on the radio and wanted to know how I was. He knew I was nervous, so he had some music put on the radio to relax me.

At 8:41, I felt a shudder of valves closing and propellant lines pulling away. At 8:51, I was told I was at the fifteen-minute mark. At 9:06, right on schedule, I heard, "Ignition," then there were vibrations, whining and rumbling sounds.

I was worried that something could go wrong, but everything was going as planned. I shouted, "*Poyekhali,*" "Let's go." I heard a whistle and an ever-increasing noise. The rocket trembled as it rose, slowly at first, then faster and faster. The noise wasn't much louder than a jet plane, though.

Two minutes into the flight, it was hard to speak into the microphone. The g-forces from the rocket's acceleration were pulling at my face muscles, but the stress was not too uncomfortable. It was like doing a tight turn in a MiG. Then, the four boosters fell away and the g-forces subsided until the sustainer stage accelerated. Again, I felt the strain from the g-forces.

At three minutes from launch, the nose fairing separation system fired, and the fairing[7] fell away allowing me to look out the portholes at the dark sky above. At five minutes, the sustainer dropped off, and one small rocket engine took over.

At nine minutes, I was in orbit. When the engine shut down, the vibration ceased, and the ship's noise became more apparent. There were fans, ventilators, pumps and valves all working to support the various on-board systems. I was travelling eastward at 28,800 kilometers per hour, but there was no sense of speed.

"How are you feeling?" asked ground control.

"The flight continues well. The machine is functioning normally. Reception excellent. Am carrying out observations of the Earth. Visibility good. I can see the clouds. I can see everything. It's beautiful!"

At seventy-nine minutes the retro-rockets fired, when I was passing over West Africa. As the automatic systems went through their paces, I thought, what will people on Earth say about my flight? My family didn't know about it until after I was in flight, and it was on the radio. I had told my wife, Valya, I was going up on 14 April so she wouldn't worry.

When the braking rockets turned on, I was no longer weightless. The g-loads pushed me into my seat, and they were much greater than at liftoff. The *Vostok* began to rotate. I was worried. I told the ground crew about it, and it soon stopped. The heat shield moved into place, but the equipment module didn't separate. The craft rotated in all different directions. It eventually slowed. I felt the temperature getting high. Was I going to burn up? I was losing consciousness from the rotation and the heat.

Finally, I was in denser air and the rotation and speed diminished. I looked outside the charred window and saw blue sky. At an altitude of seven kilometers, the hatch above me blew off. A little later I ejected. My propulsion pack and parachutes worked well, and I drifted to the ground - 108 minutes after liftoff.

I landed near the village of Smelkova[8] and was met by a group of people. I said, "Boys, let's be acquainted. I am the first spaceman in the world, Yuri Alexeyevich Gagarin." From there, I became an international hero.

And there would be more firsts for the Soviet Union by the Chief Designer, SP.

The Huntsville Times

Man Enters Space

'So Close, Yet So Far,' Sighs Cape

U. S. Had Hoped For Own Launch

Soviet Officer Orbits Globe In 5-Ton Ship

Maximum Height Reached Reported As 188 Miles

Hobbs Admits 1944 Slaying

'To Keep Up, U. S. A. Must Run Like Hell'

(From NASA Archives)

Chapter 16 - More Accomplishments

The Americans have unified their forces into a single thrust, and make no secret of their plans to dominate outer space. But we keep our plans secret even to ourselves. – Sergei Korolev

Moscow, 1961

AFTER I LANDED, I was picked up by helicopter and flown to nearby Engles airbase. There, I phoned First Secretary Nikita Khrushchev.

He said, “I’m glad to hear your voice Gagarin Alexeyevich.”

“Nikita Sergeyevich, I’m glad to report the first space flight has been successful,” I replied, using Khrushchev’s first two names.

He asked me how I felt in space, and I told him I felt fine. Khrushchev went on to say, “We shall celebrate together with all the Soviet people. Let the world look on and see what our country is capable of.”

After a medical exam, I was flown to Vnukova Airport, near Moscow. As we approached the airport, a squadron of MiGs followed us. First Secretary Khrushchev, his wife, SP and several others caught up with my wife and daughter, Valya and Anna, and met me at Vnukova. They had a brief reception there. Khrushchev embraced me. However, SP was kept to the side to keep his identity a secret.

From Vnukova, we went to Moscow. I was escorted in a long motorcade of high-ranking officials through the streets of Moscow to the Kremlin. At the Kremlin, there was a large celebration. SP was there, but, again, to keep him anonymous, he wasn't allowed to wear any of his medals.

Back at the Cosmodrome, in Baikonur, I was greeted by the entire work force in front of SP's design bureau. This became a tradition for every cosmonaut on his return from space.

Our firsts in space caused the Americans much consternation. Finally, on 25 May 1961, President John F. Kennedy declared, "I believe that this nation should commit itself to achieving the goal, before this decade is out, of landing a man on the moon and returning him safely to earth."

<><><>

LESS THAN A MONTH LATER, the American, Alan Shepard, reached an altitude of 187.5 kilometers in a ballistic sub-orbital trajectory. In July, another American, Virgil Grissom reached 190 kilometers, in a similar trajectory, but, on his return, his capsule sank. The Americans, unlike us who came down on land, landed their spacecraft in the ocean.

The Americans' sub-orbital flights were overshadowed in August, with Gherman Titov completing seventeen orbits in about twenty-five hours in *Vostok 2*. During the flight, the effects of a prolonged time in orbit and man's ability to work during an extended period of weightlessness were investigated. Titov had a movie camera with him and took ten minutes of pictures of the Earth. Like me, he experienced problems with separation of the service module after retrofiring.

Also, on that day in August, Titov's multiple orbits accomplished another first for the Soviet Union. However, he experienced severe space sickness.

One of the workers at his celebration said, "You have achieved another first for the Soviet Union. You were the first to vomit in space."

<><><>

IT TOOK A YEAR TO PREPARE for SP's next advance in the man in space program. Our military was unhappy about being required to support planetary and manned programs, which, to them, were of no value for national defense, so SP placated them by developing a spy satellite.

Kosmos 4, launched in April 1962, was the first of a series of converted *Vostoks* that would be used for photographic reconnaissance. They became known as the *Zenit*[1] program, which SP turned over to another design bureau. They had a long road ahead to develop a camera with the resolution to discern objects on the ground from Earth orbit.

About the same time, SP had one of his greatest accomplishments in the Soviet space program, the *Molinya*[2] communication satellite. He saw the need for telephone, radio and television across all of Russia and pushed for a satellite program to achieve this goal. However, he ran into tremendous opposition.

He told the higher authorities, "The further conquest of space will make it possible, for example, to create systems of satellites making daily revolutions around our planet at an altitude of some 40,000 kilometers, and to assure universal communications and the relaying of radio and television transmissions. Such an arrangement might prove more useful, economically, than the construction of radio relay systems over the whole surface of the earth. The great accuracy of movement of these satellites will provide a reliable basis for solving navigational problems." He eventually received funding for this project.

In the early 1960s, there was no rocket capable of launching a satellite to a circular geosynchronous orbit, about 36,000 kilometers above the Earth. At this distance from the Earth's surface, the satellite would travel at the same speed as the Earth rotates, causing it to stay over the same location on Earth, ideal for communication purposes.

This was a very difficult task, but SP cleverly used a four-stage version of the R-7 to put the *Molinya* spacecraft into a highly elliptical orbit of 400 by 40,000 kilometers. The higher a satellite orbits, the slower it travels, thus the upper section of the orbit took eight hours to cross the country. Three satellites with this orbit provided communication for an entire 24-hour day.

<><><>

THE *ZENIT* AND *MOLINYA* adversely affected SP's man in space funding and schedule. Nevertheless, in August 1962, *Vostok 3 and Vostok 4* were launched within a day of each other. Andrian Nikolaev completed sixty-four orbits, and Pavel Popovich followed with forty-eight orbits. The two spacecraft came within five kilometers of each other.

Both Nikolayev and Popovich spent time out of their seats each day, conducting a series of tests to determine their ability to maneuver and work in conditions of weightlessness. The two cosmonauts landed 193 kilometers and six minutes apart from each other. Again, more firsts for the Soviet Union - the most orbits, and two launches coming close together in orbit.

The following year, the next series of *Vostoks* were launched in June 1963. Valeri Bykovsky, in *Vostok 5*, passed within five kilometers of Valentina Tereshkova, in *Vostok 6*. Bykovsky set a record for the number of orbits at eighty-one, and Tereshkova was the first woman in space.

The men joked about Bykovsky, "He wasn't able to get closer than five kilometers to a beautiful woman."

Tereshkova had more serious thoughts on her time in space, "Anyone who has spent any time in space will love it for the rest of their lives. I achieved my childhood dream of the sky."

<><><>

FOLLOWING *VOSTOK*, SP planned to move forward with his *Soyuz*[3] spacecraft, which would be able to dock with other spacecraft in orbit and exchange crews.

However, Khrushchev directed him to cheaply produce more firsts for the man in space program, which included a multi-manned flight. SP resisted the idea as *Vostok* was a one-man spacecraft, and the three-man *Soyuz* was several years away. Khrushchev was not interested in technical excuses and let it be known that if SP could not do it, he would give the work to SP's rival, Vladimir Chelomei.

SP had no choice, so he had one of the cosmonauts, who was also an engineer, devise a plan to meet Khrushchev's demands. Konstantin Feoktistov determined if they didn't take their space suits and eliminated the emergency retrorockets, they could fit three men into the *Vostok* capsule. It was very risky. If the capsule leaked air, or if they needed to make an emergency landing, they wouldn't survive.

Regardless of the risks, in October 1964, *Voskhod 1* made sixteen orbits with Feoktistov, Vladimir Komarov and Boris Yegorov. It was the first three-man flight. It was simply a modified *Vostok*, but it was given the name *Voskhod 1* to make the Americans think we were further ahead in spacecraft design. The Americans' *Gemini 3*, with a two-man crew, went into orbit months later in March 1965.

<><><>

STILL WANTING TO BEAT THE AMERICANS, Khruschev called for another first. As a result, in March 1965, *Voskhod 2* made seventeen orbits. Pavel Belyaev and Alexi Leonov were the cosmonauts, this time with space suits enabling Leonov to exit the spacecraft for twenty minutes.

Leonov, tethered to *Voskhod*[4], was unable to reenter the airlock due to the stiffness of his inflated spacesuit. He had to bleed air from the suit in order to get into the airlock. After he finally managed to return to the cabin, the hatch wouldn't seal. Leonov told mission control, "The spacesuit started behaving absolutely different from what it did on the ground!"

The environmental control system compensated by flooding the cabin with oxygen, creating a serious fire hazard in a craft only qualified for sea level nitrogen-oxygen gas mixes.

Regardless of the problems with the excursion outside the vehicle, it was another first for the Soviet Union. The first American extra vehicle experience took place in *Gemini 4,* about three months later in June of 1965.

<><><>

THE DEVELOPMENT OF A VEHICLE to put a man on the Moon started in 1959, in parallel with the other programs. Initially, it was conceived as the N-1 to deliver very heavy payloads to Earth orbit or smaller payloads beyond Earth orbit, such as a Mars probe. There were various configurations studied to achieve these objectives, including several with nuclear propulsion for deep space probes.

When Kennedy announced in 1961, that an American astronaut would land on the Moon by the end of the decade, the N-1 program became focused on a lunar landing to beat the United States in this endeavor. The situation with the N-1 was more complex as SP could not come to an agreement with his old rival, Valentin Glushko.

Glushko was now the leading designer of rocket engines[5] and believed liquid oxygen and liquid hydrogen to be a dead end. SP wanted them as they were safer and more powerful. SP bypassed Glushko and got total control of the project in October 1965, when the official decree came to pursue his N1-L3.

The N1-L3 had five stages. The first three N1 stages were to carry the two upper stages into Earth orbit with two cosmonauts. The fourth stage, Block D, would fire and send the L3 vehicle out of Earth orbit toward the Moon. After separation Block D would crash onto the Moon. The fifth stage, Block E, would lower the lander out of lunar orbit and take the LK lander to the landing site. The LK lander would carry one of the cosmonauts to the surface of the Moon and back to the spacecraft. After casting off the LK lander, the spacecraft would circle the Moon with the two cosmonauts and return them to Earth.

Unfortunately, SP's health problems impacted the program. He had a medical analysis, which showed he had cancer in the large intestine. Surgery for removal of the disease was scheduled for 14 January 1966.

Alexi Leonov and I saw SP before he went into the hospital. He told us much of his life story, and I was shocked to hear how badly he had been treated in the gulag. It was such a fantastic story, and I thought it needed to be told.

<><><>

ON 6 JANUARY 1966, SP was admitted to the hospital for surgery. Unfortunately, the operation was a failure, and he died on the operating table on 14 January 1966. He was fifty-nine.

For the first time, the Soviet people learned who the Chief Designer was. *Pravda* ran a two-page obituary, and SP was given a state funeral. Thousands gathered as his ashes were carried through Red Square to the Kremlin Wall, followed by speeches by the Party leaders.

I gave the final eulogy, "The name Sergei Pavlovich is linked with a whole epoch in the history of mankind – the first flights of artificial Earth satellites, the first flights to the Moon and to the planets, flights by human beings is space, and the first emergence of a human being into free space." Beyond all of this, I felt the loss of my friend, SP, most deeply.

Epilogue

SERGEI PAVLOVICH KOROLEV is considered by many to be the greatest of the Space Age pioneers due to his ability to organize and manage so many space programs at one time, with very limited resources and to overcome many opposing political forces. In 1956 and 1961, he was given the title of "Hero of Labor." Posthumously, he was given the Lenin Award in 1971, and the Tsiolkovsky Gold Medal. To honor his memory, the Soviet Union established the S.P. Korolev Medal of the USSR. He was inducted into the International Space Hall of Fame in 1976.

<><><>

THE *SOYUZ* SPACECRAFT AND THE R-7 LAUNCHER - whose latest version is also known as *Soyuz* - remain the workhorses of the Russian space program and are still active more than 40 years after Korolev's death. With more than 1,700 flights, the *Soyuz* family of launchers still holds the record for the most launches and will probably continue to hold it for several more decades. It also holds a high mission success rate.

<><><>

YURI GAGARIN never went into space again after his first flight on 12 April 1961. He was considered a valuable national asset causing the Soviet leaders to forbid him another flight. Gagarin went back to being a test pilot and died in a MiG-15 crash on 27 March 1968. His ashes were interred in the Kremlin wall, not too far from his friend, SP.

<><><>

GHERMAN TITOV, after his first flight, continued to work for the space program. However, he was prevented from going into space after the death of Gagarin, as the Soviet authorities were concerned about losing another famous cosmonaut. Titov served in the Soviet Air Force and became General-Colonel, and in his final years, he became a Communist politician.

<><><>

ALEXI LEONOV had a long career in the Soviet space program. Leonov's second trip into space also was significant. He commanded the Soviet half of the 1975 Apollo-Soyuz mission – *Soyuz 19* - the first joint space mission between the Soviet Union and the United States. From 1976 to 1982, he was the commander of the cosmonaut team and deputy director of the Yuri Gagrin Cosmonaut Training Center. In 2006, Leonov and former American astronaut, David Scott, published a work on a dual history of the Space Race between the United States and the Soviet Union, titled, *Two Sides of the Moon: Our Story of the Cold War Space Race.*

<><><>

THE SOVIET UNION LOST THE RACE to a manned lunar landing when two Americans landed on the Moon on 20 July 1969. There were several reasons for this loss. The N1 was underfunded, lacked testing and testing facilities for large stages, and started development in October 1965 almost four years after the American's Saturn V program began. There were numerous spacecraft failures, and the whole N1 scheme was doubtful of being able to satisfy mission payload requirements.

In addition, Korolev was forced to go with much smaller engines, as Glushko, and his engine design bureau, didn't support an engine that used liquid oxygen and kerosene. Thus, Korolev used thirty RD-15s, supplied by another design bureau, on the first stage to get the required liftoff thrust for the N1 mega-vehicle. The large number of engines created a very complex system and was the cause of several failures.

Each of the four attempts to launch an N1, between 1969 and 1972, failed. During the second launch attempt on 3 July 1969, the N1 rocket crashed back onto its launch pad shortly after liftoff and exploded. It was the largest explosion in space history. After two more failures, the N1 program was shut down in 1974, and canceled in 1976.

Certainly, Korolev's death in 1966 was a contributing factor in the Soviet's loss of the Space Race to the Moon.

Although the Soviets were in a race to the Moon was always suspected, and the US may have had intelligence on the details, the race was never acknowledged by the Soviets until after the Cold War. And they even tried to hide all evidence that there ever was an N1 program.

Yuri Gagarin (left) and Sergei Korolev (right), 1961
(Photo from britannica.com)

Part Five

Wernher Von Braun

Chapter 17 - The Telescope

For my confirmation, I didn't get a watch and my first pair of long pants, like most Lutheran boys. I got a telescope. My mother thought it would make the best gift. – Wernher von Braun

Berlin, Germany 1924

"WERNHER, WHAT'S THIS ALL ABOUT?" Father asked. "You and your brother are in trouble."

"But, Father, we were just testing our rocket-powered wagon." Sigismund, my older brother, and I had bought six large skyrockets and strapped them to my wagon. We lit the fuse, and I climbed aboard. The wagon accelerated and soon was out of control. It ran into the legs of a woman walking along the street, ruining her stockings, then it crashed into a fruit stand before the rockets burned out before the wagon came to a stop. The police heard about our test, brought us to the station and notified Father.

Father said, "No more experiments. The police are releasing you to me. I'm taking the two of you home, where you will be confined for two days." Father paid the fine and paid for the damages before we left the station.

Mother was sympathetic and practical. She said, "The world needs live scientists, not dead ones."

She was bothered that I wasn't doing well in school, as I was more interested in working on a homemade automobile with a friend. I failed math and physics that year because of my non-academic interests. However, a year later my whole life changed.

<><><>

I WAS BORN 23 MARCH 1912 at Wirsitz, which was in Prussian Germany at that time. My father was Baron Magnus Alexander Maximilian von Braun, and my mother was Baroness Emmy von Quistorp von Braun. My older brother Sigismund, my younger brother Magnus and I were all barons at birth. Although we were considered nobility, we were not very high in the aristocracy.

Father had farming and banking interests and became a provincial and eventually a national government official, mainly in the agricultural field. His nobility, however, did provide him with wealth and status. We lived in very nice homes, had servants and Father had significant land holdings with workers who tended the land. Mother was also of Swedish-German aristocracy and was raised in England. She was a brilliant woman and understood people very well.

When I was born, Father thought I might become a landed gentleman and take up farming. However, he quickly saw I was different. He remarked to Mother, "He's only four and he can read the newspaper – upside down as well as right side up."

My brothers did well in school, and later on, I did, too. I showed an interest in piano, and Mother taught me to play. When Father was promoted to a national government position in 1920, we moved to Berlin, and I continued with the piano there. Father and Mother thought I might make music a career, but all that changed when I began testing our rocket-powered wagon.

<><><>

WHEN I WAS THIRTEEN, Father sent me away to boarding school, a school that was known for advanced teaching methods consisting of strong academics and hands-on experience with the trades. This was a life changing experience. Change started when I was confirmed in the Lutheran church, and Mother gave me a telescope for the occasion. It was through this gift that I became an amateur astronomer, which led to my interest in the universe, and piqued my curiosity about a vehicle, which would one day carry a man to the Moon.

At this point, I began reading science fiction. I read the works of Jules Verne and H.G. Wells, which inspired me even more. When I read the scientific treatise, *The Rocket into Interplanetary Space* by Hermann Oberth, I was surprised to find the book described space travel in terms of mathematical equations. It caused me to master these subjects, as I decided, at that point, to make space my life's work.

I persuaded Father to transfer me to a school on the North Sea. I wanted to accelerate my studies and graduate a year early. This would allow me to join a group of rocket enthusiasts in Berlin, but the headmaster preempted my plan when he had me teach classes for a teacher who had become ill.

I was able to convince the headmaster to purchase a five-inch refracting telescope as I had outgrown the one Mother gave me. I convinced some of my fellow students to build an astronomical observatory for the new device. I thought about the future often and predicted man would develop orbital flight and proceed beyond that. I put these thoughts into a five-page short story about life on a space station.

In 1929, I graduated high school with honors and was set for higher education. I knew that more education was needed for me to develop rockets, but education wouldn't put me in contact with current work in this field. Thus, in the fall, I was determined to meet my idol, Hermann Oberth. He had a group of followers, one of whom was Wiley Ley. Through Wiley, I met Oberth, who was receptive to my ideas and took an interest in me.

In the spring of 1930, at eighteen years of age, I enrolled at Charlottenburg Technical Institute, in Berlin, to study mechanical engineering. At the same time I became an apprentice to Professor

Oberth.

Oberth, in need of money for his research, said, "Wernher, help me with a department store display on interplanetary rockets."

I stood in the store for eight hours at a time raising money for rocket research. I told housewives, "An interplanetary rocket would cost 7,000 marks[1] and take a year to build." I sometimes added, "I bet you that the first man to walk on the Moon is alive today somewhere on this Earth!" Unbeknown to me at the time, Neil Armstrong was an infant when I said that.

Oberth's support group, the VfR[2], was a group consisting of students and other rocket enthusiasts. The organization's objective was to raise money for Oberth's experiments. I joined the group and soon found myself, and the others in VfR, begging for money to pursue our objectives.

As Oberth's assistant, I was deeply involved in his experiments. These rockets were very primitive and extremely dangerous. It's a wonder the ignition system didn't kill one of us. I remember opening the valves to supply propellants to the combustion chamber, then, Klaus Riedel lighted a rag soaked in gasoline, which he threw into the nozzle and ran for cover. The engine started with a roar and ran for ninety seconds until the fuel was exhausted and we were, too. In this manner, we conducted more than thirty tests, and we all survived.

Although many Germans were interested in space at that time, many thought Oberth was delusional. I thought he made a lot of sense and was fortunate to have worked with him. He was trying to prove four things: you could build a machine that could climb beyond our atmosphere, that man could leave the gravity of Earth, that man could survive in a ship in space, and that the exploration of space could be profitable. He was way ahead of the thinking of most people at that time.

As part of my education, Charlottenburg Institute assigned me for a few weeks to a machine shop in a local locomotive factory to instruct me in the practical aspects of engineering. While there, the foreman handed me a chunk of iron and said, "Make this into a perfect cube. Here are your tools." He pointed to a vise and a file.

I had trouble making perfect right angles. The foreman simply said, "Keep filing."

Five weeks of filing and five weary fingers later, the foreman looked over his glasses and simply said, "*Gut. Ja, gut!* This is what we mean," referring to the value of self-discipline and persistence in all things.

Meanwhile, Oberth and the VfR developed a small rocket engine, which was a technical success. However, funding for the project ended, and Oberth returned to his native Romania. We, the VfR, continued our work at an abandoned field outside Berlin.

About this time, I attended the Institute of Technology in Zurich, Switzerland for a few weeks. I discussed with my friend, Constantine Gonzales a medical student I met at the Institute, the details of sending a rocket to the Moon.

Gonzales suggested, "Before you attempt a lunar flight, it might be worthwhile to try it with mice as passengers."

I agreed, so we rigged a centrifuge and powered it by modifying a bicycle to test the effects of high g-forces on the mice. Some of the mice didn't survive the high forces. My friend dissected them to determine the cause of death and found the high accelerations caused cerebral hemorrhaging in the animals.

I was already convinced that we could build a machine that could climb beyond our atmosphere – Oberth's first point. Now, the experiments with mice convinced me man could withstand the rapid acceleration and deceleration of space flight if the forces were kept within limits. In addition, man could leave the gravity of Earth and survive in a ship in space – Oberth's second and third points. The profitability of space exploration was still to be determined – Oberth's fourth point.

After this work, in 1931, I returned to reenter Charlottenburg Institute to complete my studies there. I rejoined the rocket group, and we launched eighty-five rockets to altitudes of up to 365 meters by the end of the year.

<><><>

THE FOLLOWING YEAR, I received a bachelor's degree in mechanical engineering with an emphasis on aeronautics. As a means to continue my rocket work and build larger and more capable rockets, in October 1932, I went to work for the German army to develop ballistic missiles. While engaged in this work, I received a doctorate in physics from the University of Berlin in July 1934. I now had all the education credentials for my life's work.

By January 1933, things had changed when Adolf Hitler became chancellor of Germany. He suspended the constitution and ruled by decree. Germany was under the restrictions of the Treaty of Versailles that ended World War I. The treaty forbade Germany from having most weapons. But there was a loophole. The treaty did not mention rockets, which caused the German military to consider rockets as weapons.

German army ordnance, upon seeing my work with rockets, asked me to continue my research at their new Kummersdorf facilities. I had a small staff of scientists and technicians, and we eventually launched two liquid propellant rockets. This got the attention of some very influential military personnel.

Wernher von Braun carrying an experimental rocket to launch site at Raketenflugplatz, Berlin for flight-test, circa 1930. (From U.S. Space and Rocket Center)

Chapter 18 - On to Peenemünde

Man himself is the only ultimate force for good or evil.
– Wernher von Braun

Kummersdorf, Germany 1934

THE SUCCESSFUL LAUNCHES of the two A-2 rockets "Max" and "Moritz," which we humorously named after two German cartoon characters, caused the *Luftwaffe*[1] to become interested in our work. They offered us five million reichsmarks[2] to develop a rocket-powered fighter plane. The German army, not wanting to have the *Luftwaffe* get control of their project, came up with six million reichsmarks in 1935 to support our group along the current objectives.

All of this was difficult for me to believe, as I was only twenty-three years old and responsible for eleven million reichsmarks of funding. Following this windfall, million after million flowed in to support our work. Most of this money, during the next two years, was spent planning a joint army and air force rocket R&D center on the Baltic Sea.

When the German army became interested in our group, Captain Walter Dornberger became our mentor. He had helped me enroll at Friedrich-Wilhelms University to pursue my doctorate in physics. Now, in 1935, with the clandestine German military buildup well under way, Dornberger came to me one day and said, "The Ordnance Department expects us to make a field weapon capable of carrying a large warhead over a range much beyond that of artillery. We can't hope to stay in business if we keep on firing only experimental rockets."

We had already started preliminary design of our next generation rocket, the A-4. Thus, Dornberger and I continued looking for a place where we could pursue this work away from the general population. I found the perfect place at Christmas 1935 when visiting my father at his estate in Silesia. Father had quit all public offices when Hitler came to power and had retired to his estate southeast of Berlin.

That Christmas, I mentioned I was looking for a coastal site to use as a test facility.

"Why don't you look at Peenemünde?" Mother asked me. "Your grandfather used to go duck-shooting there."

I checked out the area on the Baltic Sea and fell in love with the place. Not only was it removed from the main population areas but it was a good spot for sailing. In 1937, it was selected for major R&D operations. Construction began on facilities to build and test the A-4 missile, now a high priority for the Nazi regime.

Father had warned me, back in 1933, that war lay ahead.

"But the Fuhrer wants peace," I had protested.

"I've heard that the Berlin Art Gallery now has a huge picture of Hitler in shining armor. Men don't wear armor unless they mean to fight."

Meanwhile, my team at Kummersdorf was secretly moved to the army side of Peenemünde, where Walter Dornberger, now a colonel, was the commanding officer, and I was made the technical director. The air force occupied the other side of Peenemünde for their research.

<><><>

IN 1938, I WAS IN CHARGE of the technical work, but many of the men who didn't work directly for me mistook me for a young manager. Being only twenty-six at the time, and not even looking that age, I understood how that happened. In fact, when we went to nightclubs during trips to Berlin, I was often asked to show my ID to prove I was twenty-one, the legal drinking age.

Another problem that I encountered was a reputation for being a heavy spender for rocket development work. Gold-plated parts and expensive wind tunnels contributed to this image. However, we lost some funding before World War II broke out. We thought, with the invasion of Poland in September 1939 and the declaration of war by the Allies, we would be funded more. However, in February 1940, Hitler cancelled all weapons development programs, which would not produce equipment for use within one year.

About this time, I brought Hermann Oberth to Peenemünde, where he conducted theoretical research. Moreover, he provided moral support for our work.

At that time, I had a sport plane at my disposal, and later on, I had a civilian version of the Me-109 fighter aircraft for my use. I would take one of these planes for business trips to universities, industries and other military sites, but on weekends, I would visit my girlfriend in Berlin – a much better use of the planes in my opinion. I had learned to fly in the mid-thirties as part of fulfilling my military service requirements. Fortunately, my service was part time, and I was able to continue my rocket work.

With residual funding, we continued developing and solving problems with the A-4. In 1942, we had two flight attempts, which failed on the launch pad. I could see the men were in despair, so I simply told them, "Well boys, back to the drawing board." I remembered the lesson I learned while filing the metal cube at the locomotive factory - persistence pays off.

Finally, on 3 October 1942, we had a successful flight. It reached an altitude of eighty kilometers. Everyone wept for joy. One of the men later asked me, "Did you weep, too?"

I didn't remember whether I wept, but I remembered having quite a few drinks at the party after the flight. I answered, "I don't know about my getting wet *outside*, but I got very wet inside."

The day of the flight, Colonel Dornberger said to me, and others "Do you know what we accomplished today? This afternoon, the *spaceship* has been born." He continued, "I warn you that our headaches are by no means over – they are just beginning!" He added the team must perfect the rocket as a weapon. Space travel and exploration would have to wait.

In late 1942, it was unsafe to talk of space travel as Germany was struggling to survive the war. One day, I visited the accelerometer development laboratory at the base. On leaving, a lab technician asked me, "Do you believe that someday rockets would put artificial moons in Earth orbit, as Oberth and others had predicted?"

"Absolutely. If our team stays together after the war, we might have the opportunity to put up Earth satellites, rocket men to the Moon and later to Mars." I stopped and warned, "This is dangerous talk. It's not safe to discuss these things." I quickly left the lab to avoid having to answer more questions about satellites and space travel, as the High Command wouldn't permit such talk.

There were many tests to make the A-4 a reliable rocket. During this process, one rocket exploded shortly after liftoff, and parts of it landed near us. I remarked, "I'd have been safer in the target area!" To remain positive, I added, "Now all we need to do is shift the impact point 180 kilometers farther north, and all our problems will be solved!"

Hitler was not a proponent of rockets, as he had a dream indicating rockets were no good. This, of course, hindered our progress. However, early on Field Marshall Walther von Brauchitsch, the army's supreme commander, came to our aid. He believed in us and what we were doing, so, he assigned thirty-five hundred officers and enlisted men to Peenemünde. He used the excuse they needed training, when, in fact, they were engineers, technicians and scientists.

I met Hitler four times. When we met in 1934, he seemed a pretty dowdy type. Later, I began to see the shape of the man – his brilliance, the tremendous force of his personality. At my last meetings with him, Hitler suddenly struck me as an unreligious man, a man who did not feel that he was answerable to anyone, that there was no God for him.

When he first visited our R&D facility in March 1939, Dornberger warned me, "Do not utter a word about space travel, talk only of weaponry."

Even when Hitler brought up the subject of space travel, I heeded Dornberger's warning and said nothing. Later, in a briefing to him and the top brass, I never broached the subject.

After the meeting, I told my colleagues, "Hitler seemed interested and pleased with our work."

One colleague asked, "How do you remain so calm with such a formidable audience?"

"I just use my imagination."

"How do you mean?" someone asked.

"Well," I answered, "you simply imagine that all those guys sitting out there in front of you are dressed only in their underwear – and then you lose all your awe of them"

I was forced to join the Nazis, however, when one of Himmler's assistants offered me a commission as a second lieutenant in the *Schutzstaffel*, the SS.

I tried to avoid the commission by saying, "I'm so busy with my rocket work that I have no time for any political activity."

The assistant persisted, so I went to Colonel Dornberger for advice. "Rejecting the commission is dangerous. It would be taken as disloyalty and a personal insult by Himmler." [3]

I accepted, and, fortunately, they required only some minor activities, such as training personnel to spot enemy planes over the base.

In 1943, as the war was shifting against Germany, Hitler became interested in the A-4 development and production. He realized a large number of workers were needed to mass-produce the rocket. As a result, he turned the production over to the SS. I wasn't involved in the SS recruitment of workers, which were mainly from the concentration camps. More lives were lost producing the V-2, the new name for the A-4, than were killed by it in the war. The conditions at the production site were brutal with many deaths from starvation and overwork. There was nothing I could do, as the SS was in charge.

In February 1944, I was summoned to Himmler's headquarters in East Prussia. I didn't know if I was going to receive a medal around my neck or a bullet in the head. I entered his office with considerable trepidation. It must be said, though, that he is really as mild-mannered a villain as ever cut a throat, for he was quite polite and rather resembled a country schoolteacher.

Himmler wanted me to join his staff, "I can do a lot more for you, Wernher, more than those stuffy Army generals."

I politely declined.

He then tried to get me to speed up V-2 production on an assembly-line basis.

"You can't push too fast on some things," I told him. Remembering that he was interested in horticulture, I said, "It is all right to put a plant in the ground and carefully nurture it, but too much manure could kill it."

Himmler didn't receive my metaphor very well. He was obviously annoyed. I didn't think much of his reaction until I was arrested by the *Gestapo*[4] three weeks later.

At a *Gestapo* prison, I was accused of sabotaging the war effort by having my heart set on space rather than destroying London. This was considered high treason.

Dornberger, now a major general, stepped in for me. He and Albert Speer,[5] an enemy of Himmler, went directly to Hitler. They told the dictator that the V-2 program would falter without me. Hitler complied with their request to release me, and I was put on probation. I wasn't ill-treated in prison, in fact, it gave me a chance to rest and to think.

Even after production began on the V-2, its systems were not completely ready for mass production. It took more testing and refinement to make the rocket reliable. Beginning in September of 1944, the V-2s, with a range of 320 kilometers, were launched against Paris, London, Antwerp, and other targets from mobile platforms.

As the Russian Army advanced on Peenemünde in 1945, it became apparent that things were coming to an end. We were now under the direct command of the SS, and we feared that SS General Hans Kammler[6] might possibly use us as a bargaining chip or have us executed to keep us from being captured by the Allies.

I received several contradictory orders from German High Command, which was in mass confusion at the time. I had ten orders on my desk. Five promised death by firing squad if we moved, and five said I'd be shot if we didn't move. Since I was damned either way, I called a meeting in mid-January 1945 with the other top officials at Peenemünde. The rumor was we would either be killed in combat or taken prisoner by the Russians. Either of those prospects was very grim to us.

We, except for Helmut Gröttrup who later went to the Russians, decided to surrender to the Americans. We were more likely to be able to continue our research after the war in the United States. We knew that somehow, we had to smuggle all our research papers and important equipment out of Peenemünde. We needed a plan to prevent our work from being destroyed or falling into the wrong hands.

Launch of a V-2 (Photo from German Federal Archives)

Chapter 19 – Operation Paperclip

Research is what I'm doing when I don't know what I'm doing.
– Wernher von Braun

Peenemünde, early 1945

FINALLY, IN FEBRUARY 1945, orders came for us to move south to the Harz Mountains. This was good for us as the Americans were coming in from that direction. To ensure we could move our equipment and all the supporting technical papers, I forged another set of orders. On official SS paper, I created instructions to move V-2s, supporting equipment, and technical specifications and documentation with us. I labeled all of this VABV, Project for Special Dispositions.[1] I had all of our trucks and other vehicles marked VABV, as well.

I confidently showed our "orders" to our security guards and the guards we encountered along the way, hinting we were performing a top-secret operation. We were waved on at each stop. I stopped at a Baltic farm where I met my uncle, Baron Alexander von Quistorp, and his lovely fifteen-year-old daughter, Baroness Maria Louise. Later on, she would become a major part of my life.

We traveled by night to avoid being seen by enemy aircraft. Along the way, my driver lost control of our vehicle and went off the road into a ditch. Sadly, he was killed instantly. My left upper arm was badly broken and required a large cast to set the bones. Later, when I met with the Americans, I was still in the cast.

Finally, we stopped at Bleicherode, near Mittewerk where the underground missile factory was located. We settled into empty factories and other abandoned buildings and hid tons of V-2 equipment and documents in caves and old mines.

In March, we received orders from SS General Kammler to move to Oberammergau in the Bavarian mountains. We were ordered to destroy all of the secret documents we were previously told to take with us. Instead, we hid them in old mines and dynamited shut the mine entrances before leaving. The Americans later retrieved these documents.

When we arrived at Oberammergau, we were kept behind barbed-wire fences and were under guard. However, we were able to convince the guards that this was a very dangerous situation that an enemy air raid could kill all of us in one attack. They listened and allowed us to disperse to surrounding villages in early April 1945. On April 30th, we heard that Hitler had died. When his death was made public a few days later, our SS guards began dispersing enabling us to send my brother Magnus, to find the Americans. Magnus was selected as he spoke better English than the rest of us.

Magnus indeed found the Americans, an U.S. Army antitank patrol. He told them, "We are a group of rocket specialists up in the mountains. We want to see your commander and surrender to the Americans." Soon after, we met with the Americans. Thus began a long friendship that brought us to America and ultimately to Fort Bliss, near El Paso, Texas.

<><><>

THERE WERE 127 OF US who went to Fort Bliss, under Operation Paperclip, leaving behind several hundred others the Americans wouldn't take. We weren't permitted to bring our families as we were initially under a six-month contract to assemble and test the V-2s that went to America, too. As the Americans saw how much we had accomplished with the V-2, they extended our contract and eventually made us part of the U.S. Army team developing rockets. In the process, our families later joined us.

On the train to Fort Bliss, in September 1945, I had a close encounter with one Texan. He sat next to me on the train, and, being a friendly sort, asked me where I was from.

Not wanting to reveal my identity and have to explain what I was doing in America, I simply answered, “Switzerland.”

“Oh, yes. I’ve traveled all over Switzerland. What’s your business?” he asked.

“Uh, the steel business.” I improvised in my limited English.

“Why, that’s mine, too!” the Texan exclaimed. “What’s your product?”

“Ball bearings,” I answered, confident the conversation would stop there.

“That’s mine, too.” He said again.

Now, I was in trouble. Not having much to do with Swiss ball bearing manufacturers, he would begin to ask questions and even suspect I was a German spy.

At that point, the train stopped, and the Texan jumped up to leave. Before he left, he shook my hand, slapped me on the back and said, “If it hadn’t been for you Swiss, I doubt that we could have beaten those Germans!”

I was amused by this remark, but more relieved, as our presence in America was being kept secret that first year.

Probably what impressed me the most during my trip from the east coast to Texas was the massive number of automobiles, especially in the big cities. It still startles me to be able to travel for days in America without crossing borders or being subject to passport or custom control, such as you constantly encounter in Europe.

<><><>

THE REST OF OUR TEAM ARRIVED in the ensuing months. They were housed in surplus barracks at Fort Bliss, and I was initially sent to an army hospital as I had contracted hepatitis. At that time, no one knew of our mission, except for the president and Colonel Toftoy at the Pentagon. Toftoy assigned Major James Hamill to be our overseer with me as the project director.

At first, the team was confined to the base with their family members back in Germany. Most kept active physically by playing sports, while others gardened or made furniture. Some created a clubhouse from an old shack, which was used for weekend parties. After a while we were allowed to go to the movies, a restaurant, or go shopping in El Paso under an army escort. We were permitted to go town once a week.

We worked with the U.S. Army and civilians, mainly from The General Electric Company whom the Army had hired during the war. We trained them in the assembly and launch of the V-2, which had arrived in August.

No V-2 arrived in launch-ready condition. Had an assembled missile been received, it would have been disassembled so individual components and subassemblies could be tested properly. Our previous experience proved that a large increase in flight failures resulted when assembled missiles were stored prior to use. These failures were minimized by testing each component prior to installation in the vehicle, and by checking various connections for leaks and electrical disconnects. Therefore, all missiles launched from White Sands were assembled at the missile range from basic components and parts brought from Europe and tested under our supervision.

We had our first launch on April 16, 1946, at White Sands Missile Range in New Mexico. The launch went well initially, but suddenly the rocket began to tumble and ultimately crashed. The failure was traced to a jet vane[2] falling off shortly after launch. I told the responsible engineer, who was with me, "Herr Steurer, that was one of your lousy jet vanes! You materials people have to do everything in a destructive way!"

The following month, however, we had a successful launch. With a roar, it inched from the launch pad, hovered in space for an instant, then, with a sudden lurch, climbed into the sky at supersonic speed.

I began improving my English soon after I arrived at Fort Bliss. By talking to army personnel and studying books on the subject, within a few months, I could speak the language, although I definitely had a German accent. When I was able to converse fairly well, I was required to make various presentations on our work. During one of my first presentations, I noticed some scowls on the faces of my audience. I didn't understand the reason behind the strange looks, but when I continued the looks continued, too.

After the meeting, an Army official came to me and said, "Doctor, I suggest you understand American slang before your next presentation."

I got the message, and, apparently, the army had gotten a message they didn't expect.

<><><>

IT WASN'T UNTIL DECEMBER 1946 that our presence was made public knowledge, although, the previous May, Major Hamill announced to Congress our team was at White Sands working on V-2s. The December 4th issue of the *El Paso Times* gave us a front-page story. As a result, we generally were well treated by the local citizens, but we did occasionally encounter some negativity toward us. I encouraged the team to keep their spirits up, which contributed to us being accepted by most people.

In January 1947, I was invited to speak at the El Paso Rotary Club. My topic was "The Future Development of Rocketry." I began my speech by addressing the issue that rockets started as a weapon, "It seems to be a law of nature that all novel technical inventions that have a future for civilian use start out as weapons." I continued with my vision of Earth satellites, astronauts, trips to the Moon and beyond, and manned space stations. My talk was well received, and the audience gave me a standing ovation when I finished.

The next month, I received army approval to return to Germany. I had sent a letter to my father to determine *discreetly* if my cousin, Maria Louis von Quistorp, would marry me. My father, with the subtlety of a bulldozer, waved the letter in front of her and said, "I'm supposed to find out if you will marry Wernher. What should I tell him?"

Fortunately. Maria wrote me, "I told him that I'd never thought of marrying anyone else."

I was almost thirty-five, and she was eighteen when we married on March 1, 1947. A marriage to a cousin and such an age difference were not unusual in Europe at that time, although it shocked some Americans.

Our honeymoon was somewhat unusual as two U.S. military policemen stayed with us full time to prevent the "other side" from abducting us. When we returned to Fort Bliss, we weren't without companionship, as the Pentagon had previously allowed our families to come to the U.S.

<><><>

WE CONTINUED LAUNCHING V-2S with varying degrees of success. Besides improving the rocket itself, we launched it with scientific instruments to study the upper atmosphere. One interesting application was the Bumper Program, which put a U.S. WAC Corporal rocket[3] on top of our V-2.

The Bumper Program was conceived by Colonel Holger Toftoy to investigate launching techniques for a two-stage missile and separation of the two stages at high speed. It was also to conduct limited investigation of high-speed, high-altitude flight dynamics, and to attain velocities and altitudes higher than ever reached before. Bumper was a joint venture between General Electric, the Jet Propulsion Lab, Douglas Aircraft, and us, with GE having overall responsibility.

The first launch, on May 13, 1948, reached an altitude of seventy-nine miles, but an early cut-off of the WAC Corporal second stage ended the flight prematurely. There were five other flights at White Sands through April 1949, with only one success on February 24, 1949.

On that date, the two-stage vehicle achieved a record altitude of 244 miles and a record maximum speed of 5,150 miles per hour. It was launched with a monkey on-board to understand biological effects at high altitudes and speeds. It was the first launch with an animal.

The rockets had a range that could endanger some of the population centers, which resulted in Cape Canaveral, with future launches over the ocean, being selected for future rocket work. I'm sure a wayward V-2 landing in Juarez, Mexico had something to do with the Army's decision to launch over the ocean.[4]

Bumper 8 was launched from the Cape on July 24, 1950, and Bumper 7[5] a week later on July 29th. Both launches were successful low-angle atmospheric flights reaching an altitude of ten miles and a distance of 200 miles. These were the first of many, many launches from Cape Canaveral.

<><><>

BY JUNE 30, 1951, a total of sixty-seven V-2s were sent aloft over the New Mexico desert in a program called Operation Fire-Ball. We renamed the program Operation Icebox because the Americans were keeping us on ice, as they didn't have any long-range plans for us.

The standard V-2 measured forty-six feet in length, with a thrust of fifty-five thousand pounds and a payload of sixteen hundred pounds.

Sixty-eight percent of the V-2 flights were considered successful. However, much valuable information was gained from flights with malfunctions and failures. The failures were almost equally divided between guidance system[6] and propulsion anomalies. These systems were greatly improved during the program. In addition, we developed a telemetry system, which allowed us to monitor the performance of various vehicle systems in flight. All of this resulted in a more reliable V-2 with more payload capability, and personnel trained to handle and launch large rockets.

<><><>

LIFE AT FORT BLISS was less than blissful. Our accommodations were Spartan and the food was mediocre. The men complained, and I needed to encourage them that things would eventually improve. I was discouraged, too. There was much bureaucracy, and several times I thought about quitting and working in industry. However, I knew that the chances of working on advanced projects were more likely to occur in the military due to the prevailing interest in rockets as weapons and the funding to support this work.

On the positive side, our first child, Iris Careen, was born at Fort Bliss on December 9, 1948.

In 1950, things changed when we received orders for our team to move to the Redstone Arsenal at Huntsville, Alabama. This was good news, and I told one of the men, "We are going to make history."

Bumper 8 launch at White Sands Missile Range, July 24, 1949. (Photo in the Public Domain from US Army/NASA)

Chapter 20 - Project Jupiter

The United States has no time to lose if it wants to be the first in orbit. – Wernher von Braun

Huntstville, Alabama, 1952

IN THE EARLY 1950S, we moved to Huntville, Alabama. We Germans enjoyed the move, as there were trees and hills there, unlike the El Paso area.

Our project was generally well received by the local citizens, as the town had lost a major contract from the Air Force to build a huge wind tunnel there. Tullhoma, Tennessee was the site selected due to political clout of two senators from that state.[1] As a result, Huntsville was glad to see us from a jobs perspective, since, during the war, the Army employed more than 14,000 citizens to produce "chemical munitions" there. Now, virtually no jobs existed at the Redstone Arsenal. However, the natives were as skeptical about us as we were curious about them.

There was minimal animosity, although one townsman was reported to have said, "The last time our boys had seen Germans, they were shooting at them."

The citizens listened to lectures I gave on rocketry, some believed we could go to the Moon someday, and others were very skeptical. One man who painted our house described me to one of his friends as "a nut who wants to go to the Moon."
Most of us eventually bought homes on a knoll nicknamed

"Sauerkraut Hill," or on Monte Sano, a small mountain east of town. We bought a three-bedroom house a few blocks from downtown.

I encouraged the team to become active in the community and to learn the language. When our second child, Margrit Cecile, was born on May 8, 1952, I encouraged Maria not to be a *hausfrau*, but be active outside the house. As a result, she earned a pilot's license, attended horse shows and enjoyed water sports. She loved to travel. We went to New York once a year to see a Broadway show, go to symphonies and art museums, and shop for Christmas presents. On vacations, I did my best not to talk about rockets and space travel.

Our project was what became known as the Redstone missile, a derivative of the V-2. It was originally designed to carry a 6,500-pound nuclear warhead about 175 miles. The Redstone was first flight tested at Cape Canaveral starting in 1953. My first visit to the Cape reminded me of the isolation and desolation at Peenemünde. As expected, some tests were successful and some were not.

There was a significant difference between launching from White Sands and Cape Canaveral. White Sands was a dry desert, while the Cape was next to the ocean in a hot and humid climate. This resulted in corrosion issues in the entire vehicle. We created a special team to study the effects of corrosion and determine how to deal with it.

I felt it was important to get out into the shops, labs and test sites to get a complete understanding of how the project was going and to get to know as many of the men by name as possible. Once, after one engineer briefed me on the progress in his area, I told him, "I'm just a plumber myself." With that, we both had a good laugh, as much of the work required making parts to very tight tolerances more so than a plumber would have to do.

I spent considerable time developing a workforce that was highly capable and completely honest in their work. One launch from the

Cape resulted in the vehicle going off course halfway through its flight. We couldn't determine the cause of the failure but had narrowed it down to one section of the vehicle. Various tests were conducted on this equipment prior to its installation, which showed the component was functioning correctly. Finally, we made an educated guess as to the source of the vehicle failure and corrective measures were ordered.

At this point, an engineer, who was a member of the firing group, called and said he wanted to see me. He came up to my office and told me that during pre-launching preparation he had tightened a certain connection just to make sure that there would be good contact.

While so doing, he had touched a contact with a screwdriver and drawn a spark. Since the system checked out well after this incident, he hadn't paid any attention to the matter. But now that everyone was talking about a possible failure in that particular apparatus, he just wanted to tell me the story for whatever it was worth. A quick study indicated that here was the answer. Needless to say, the "remedial action" was called off, and no design changes were required. However, the firing team was instructed not to over-tighten connections and not use a screwdriver where it may cause sparking.

I sent the engineer a bottle of champagne, because I wanted everybody to know that honesty pays off, even if someone may run the risk of incriminating himself. Absolute honesty is something you simply cannot dispense with in a team effort as difficult as that of missile development.

The Redstone made great strides in becoming "Old Reliable." Even though it became operational in the mid-fifties, many "mod kits" were still issued to the soldiers in the field. The kits continued and a meeting was held to deal with all the changes. Redstone production chief, John Goodrum, and other military personnel attended. One of the rocket team members, Arthur Rudolph, was also there and defended the changes as required to assure increased reliability.

At that point, I reached over and put my hand on his shoulder and said, “Art, you remind me of a fellow who married a virgin and kept telling her how good it was going to be once he got it perfected.”

A design freeze occurred shortly thereafter.

<><><>

ON APRIL 14, 1955, Maria and I and more than one hundred others renounced our German citizenship and became naturalized American citizens in the Huntsville High School auditorium. I told reporters, “This is one of the proudest and most significant days of my life – almost like getting married. I am very, very happy.”

Major General Toftoy was there and told us, “In the ten years since I first met you in Germany and promised you no future, this great team of rocket engineers and scientists has achieved much. I am sure that the future holds even greater things for you.”

And that it did.

<><><>

ALL THROUGH THE FIFTIES, I developed relationships with journalists and other communicators including Arthur C. Clarke, broadcaster Walter Cronkite, Walt and Roy Disney and, later on, television’s Hugh Downs.

The first significant breakthrough in spreading news of our mission came when *Collier’s* magazine picked up our cause. The head of the Astronomy Department at Harvard, Fred Whipple, California physicist Joseph Kaplan, science writer Wiley Ley, some others and I wrote a series of space articles for the magazine. In addition, I wrote several books on the subject: *Across the Space Frontier* in 1952 and *Conquest of the Moon* in 1953.

Not everyone was supportive of me promoting space in this way.

Adolph Thiel, a veteran of the V-2 days, left the program because of my methods. Later, we talked about his going to TRW Corporation in California. I told him, "We can dream about rockets and the Moon till hell freezes over. Unless the people understand it, and the man who pays the bills is behind it, no dice. You worry about your damned calculations, and I'll talk to the people."

I followed my *Collier's* articles with a series of Walt Disney television programs on space, which began airing in 1955. The shows featured me, my chief scientist, Ernst Stuhlinger, and other experts and were introduce by Walt Disney himself. *Man in Space*, *Man and the Moon* and *Mars and Beyond* were shown which helped educate the American public on space.

<><><>

IN EARLY 1956, we were put under a new command, the Army Ballistic Missile Agency, the ABMA, which was headed up by Major General John Medaris. He gave us a pep talk and added, "I will be available to fix all kinds of problems. If there are any problems you can't resolve, you let *me* know."

I didn't let him get away with a statement like that, and I quickly responded, "Ah, yes, General, there is a problem that we've been working on. On the date you have selected for the next launch, there is a full Moon, and this will make optical tracking of the vehicle very difficult."

He was quick in his response, "What are you going to do about that full Moon?"

Medaris was good about my challenge, and we joked about it later. "The Big M" and I became very good friends. He headed ABMA for four years.

Medaris was a great leader, but the U.S. government was not willing to undertake space projects. I told a journalist in 1956, "Galileo, the Wright Brothers, and Thomas Edison wouldn't have a Chinaman's chance here today. They'd be thrown right out of the

Pentagon on their ears!" At that time, many in the military considered me a nut, but things would soon change.

In the mid-fifties, International Geophysical Year, IGY, was being planned for 1957 and 1958. As part of the study of the Earth and beyond, we, and others in the military and the scientific community, began thinking of launching Earth satellites to study the environment in outer space. This eventually led to an army-navy team proposal to put up a five-pound payload to an altitude of at least two hundred miles between the fall of 1956 and November 1957. The plan was eventually called Project Orbiter.

However, instead of using a vehicle designed by Project Orbiter, the Eisenhower administration chose the Naval Research Laboratory's Project Vanguard to launch the first satellite in 1957 in support of IGY. The rocket would consist of a General Electric Viking liquid propellant first-stage, an Aerojet liquid propellant second-stage and another contractor's solid propellant third stage. It was to launch a forty-pound payload to a three-hundred-mile-high orbit.

The White House had rejected a military missile, such as our proven Redstone, in favor of a "scientific" rocket like the Vanguard even though the Vanguard was untested and unproven.

I protested this decision, "This is not a design contest. It is a contest to get a satellite into orbit, and we are way ahead on this."

The administration replied, "Vanguard possesses more dignity than your proposal." They wanted to project to the world that the U.S. was launching a rocket for peaceful purposes, not using a ballistic missile to advance our military capability under IGY.

"I'm all for dignity," I told them. "But this is a Cold War tool. How dignified would our position really be if a man-made star of unknown origin suddenly appeared in our skies?"

The Soviet Union announced that it also would launch a satellite during IGY. We later learned they were amazed that Washington had selected an unproven vehicle over our proven, reliable one.

"The United States has no time to lose if it wants to be first in orbit," I told a reporter in 1956. I warned that Russia was working hard to launch a satellite.

There was much opposition to our work back then, to the extent we were ordered by the administration to limit our work to weapons development and not work on putting up a satellite. However, the Army team pressed on with Project Jupiter, a vehicle based on our Redstone.

It was our task to compare the technical characteristics of the Air Force's Thor with our Jupiter vehicle. This allowed us to build twelve Jupiter-Cs, which were upgraded versions of an intermediate range ballistic missile we had worked on with the Navy under Project Orbiter. We didn't launch all of them for these tests but kept some in reserve just in case we were called on to launch a satellite.

One interesting development of this work was the successful demonstration of an ablative nosecone for reentry, which would prove very useful in the years ahead. For this, I received the Civilian Service Award, the Pentagon's highest nonmilitary award.

In 1956, Vanguard testing began, and it was in trouble from the start. We offered to help the Navy, even having them use our vehicle with the Vanguard name on it, but our offer was rejected. In November, it became apparent that the Soviets were firing missiles capable of launching a satellite.

In the fall of 1957, we became aware that the Soviets were on the verge of launching a satellite. Then on October 4th, what I had warned about appeared over our skies. The Soviet Sputnik satellite was passing over us. I heard about it at a dinner party, and all I could say was, "I'll be damned."

Sputnik shocked America. It struck fear into many people from the thought that the Soviets could control space. The Eisenhower administration at first dismissed its importance.

However, Texas Senator Lyndon B. Johnson told Congress, "The Roman Empire controlled the world because it could build roads. Later, when men moved to the sea, the British Empire was dominant because it had ships. Now the Communists have established a foothold in outer space. It is not very reassuring to be told that next year we will put a better satellite into the air. Perhaps it will even have chrome trim and automatic windshield wipers."

After *Sputnik II* was launched in November, with a dog on board, we were commissioned to go ahead with a launch of an Earth satellite. It was a good thing, as Vanguard failed to launch in December 1957, rising only about four feet above the launch pad.

Thus, we went ahead and launched on January 31, 1958, using a Jupiter-C vehicle. The Jupiter-C, more accurately the Juno I, was a four-stage vehicle with the Redstone as the first-stage and solid propellant rockets for the upper stages.[2] It successfully put *Explorer I* into orbit, America's first artificial Earth satellite. Much information was gained from the *Explorer* launches, which was invaluable in understanding the environment of outer space from both a scientific perspective and for future launches.

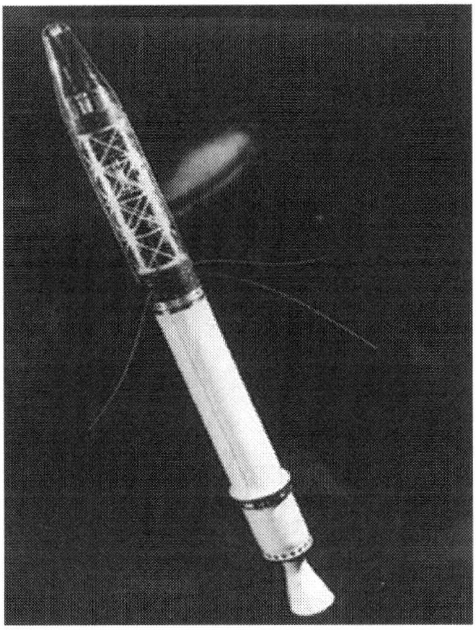

Explorer I satellite, launched on Juno 1 rocket January 31, 1958. (From NASA website)

Chapter 21 - Projects Mercury and Gemini

One good test is worth a thousand expert opinions.
– Wernher von Braun

Huntsville, January 1958

ALTHOUGH *EXPLORER I* FOLLOWED *SPUTNIK I* by several months, it is credited with having started the Space Race. However, it was much more than that. *Explorer I,* a thirty-one-pound satellite, discovered the radiation belt above the Earth, which became known as the Van Allen belts. It was named after Professor James Van Allen of the University of Iowa, who predicted its existence and designed the equipment to measure it. The belt consists of charged particles from the sun trapped in the Earth's magnetic field,[1] which protects the Earth from these high-energy particles. *Explorer III*, launched in March, mapped out the belt in more detail.[2]

There was still some complacency after the Soviets launched *Sputnik I and II.* However, when they put up *Sputnik III*, a twenty-nine-hundred-pound satellite, in May 1958, things changed quickly. In mid-July, Congress passed the Space Act, which created the National Aeronautical and Space Administration, or simply NASA. The National Advisory Committee for Aeronautics, NACA, located in Langley, Virginia, came under NASA as a result of the Space Act. There was some discussion to move our 4,800-man team to the Washington area, but General Medaris and I fought it, which resulted in us remaining at Huntsville.

Prior to Apollo, there was much rivalry between the NASA agency at Langley and ours in Huntsville. On one occasion, I took a team from Huntsville to Langley, to develop plans for Project Mercury, our first manned flight. At Langley, which I named "the first Houston" because many of the scientists and engineers at Langley later went to Houston to form the NASA Lyndon B. Johnson Space Center, I confronted NASA's Bob Gilruth.

I asked Gilruth, "I hear reports that you are planning some two-man capsule, and you're talking to the Air Force about it. Would you care to give us a briefing on it?"

"No, I wouldn't," was his curt answer.

"Well, we'll see about that," I replied, and the meeting quickly ended.

It was actually the Gemini program they were working on, and they were trying to exclude Huntsville. I went to NASA headquarters and got the assignment for us to do the vehicle part of the program.

In early March 1959, we launched Juno II, also a derivative of the Jupiter-C rocket. It carried *Pioneer 4* spacecraft in a lunar fly-by. It didn't come close enough to the Moon to determine its radiation environment, but it became the first U.S. satellite to go into an orbit around the Sun. Later launches would define the Moon's radiation exposure from cosmic rays and solar flares.

Later in 1959, we sent two small monkeys into space aboard a Juno I. "Abel" and "Baker" survived the suborbital flight and provided us with valuable information for our man in space program. Abel died shortly after the flight while Navy surgeons removed the biomedical probes. Baker lived a long life and was featured in shows at the U.S. Space & Rocket Center in Huntsville.

<><><>

IN JULY 1960, our team was officially transferred to NASA,

although we remained at Huntsville. In September, President Eisenhower came to Huntsville and formally dedicated our facility as the George C. Marshall Space Flight Center, named after the great military leader of World War II. Recognizing the need for large rockets in the future, we were working on the Saturn I launch vehicle at that time, and the president toured the project. Surprisingly, he said to me, "They come into my office and say it has eight engines. I didn't know if they put one on top of the other or what!" Clearly, he wasn't receiving much of a briefing on our work. A few months later, John F. Kennedy became president and would transform the space program into a national project. Only the building of the Panama Canal and the Manhattan Project rivaled the program's size.

We were very busy at that time in preparation for Project Mercury where we would put a one-man capsule into space. We launched a suborbital unmanned capsule in December on a Redstone missile, then one in January 1961 with the chimpanzee "Ham" on board, followed by another in late March with the chimp "Enos" in the capsule. We now were ready to launch a man in a suborbital flight.

However, on April 12, 1961, the Soviets again put us in catch up mode by launching Yuri Gagarin not only into space, but into a complete orbit around the Earth. Soviet Premier Nikita Khrushchev told Gagarin, "Let the capitalist countries catch up with our country!"

I congratulated the Soviets, and I called the flight, "the shot heard around the world." For the sake of the White House and Congress, I told reporters, "We are going to have to run like hell to catch up!"

President Kennedy responded, "It is going to take some time. We are behind." He then asked Vice President Lyndon Johnson to recommend a national space project that could produce "dramatic results" in competition with the Soviets, one that the U.S. "could win."

Lyndon Johnson, Marshall Center Director Bob Gilruth, Air Force General Bernie Schriever and I recommended putting a man on the

Moon and returning him safely. We told Kennedy, "We have an excellent chance of doing this before 1970, ahead of the Russians."

As a result, President Kennedy, on May 25, 1961, went before Congress and challenged the nation to "commit itself to achieving the goal, before this decade is out, of landing a man on the Moon and returning him safely." This became Project Apollo.

I publicly announced, "Project Apollo puts the program into focus. Everyone knows what the Moon is, what this decade is, what it means to get some people there – and everyone knows a live astronaut from one who isn't."

Even before President Kennedy's challenge, we quickly responded to the Soviets putting a man in orbit with Project Mercury. We did this in two steps. The first step was a suborbital flight on May 5, 1961, using a modified Redstone rocket.[3] Alan Shepard, in a 4,000-pound Mercury capsule, went 116 miles up from Cape Canaveral and splashed down in the Atlantic.

We had a celebration at Huntsville's Courthouse Square where I told the crowd, "Our opponents across the ocean, behind the Iron Curtain, thought about a month ago they had slammed the door to the universe in our face, but Shepard has led us out of our dilemma and embarrassment. We will go farther and farther, eventually landing on the Moon."

Gus Grissom duplicated Shepard's flight on July 21, 1961. However, the space capsule sank shortly after splashdown. Both flights lasted just over fifteen minutes.

The second step was John Glenn's three orbits around the Earth on February 20, 1962. The Redstone didn't have the capability to put a 3,000-pound spacecraft into orbit, so the Air Force's Atlas ICBM was used as the launch vehicle. The Atlas had been refined to make it more reliable for a manned launch, including an automatic abort system that would be activated if its sensors detected a vehicle malfunction.

The six manned flights in Project Mercury, ending with Gordon Coopers' thirty-four-hour flight on May 15th and 16,th 1963, showed that a man can function ably as a pilot-engineer-experimenter without undesirable reactions or deteriorations of normal body functions for extended periods of weightlessness.

Project Gemini followed Mercury in the years 1964 through 1966. The Gemini space capsule had a launch weight varying from 7,100 to 8,350 pounds. The project ran concurrent with the beginning of the Apollo Program. Project Gemini, named after the twin constellation, was headed up by Bob Gilruth of the Houston center.

In 1964 and 1965, two Gemini missions were flown without crews to test various systems and the new heat shield. These were followed by ten two-man flights in 1965 and 1966. All were with what was the most powerful rocket in America's inventory. An improved version of the Air Force's Titan I, Titan II first flew successfully in March 1962. Unlike Project Mercury's Redstones and Atlases, as well as the later Saturn Vs, the Titans used room-temperature propellants called hypergolics, which ignite on contact.[4] The fumes are highly toxic. Gemini, like Mercury, gave us additional knowledge on spaceflight, which included the astronauts' reactions and observations during each mission.

Gus Grissom and John Young were the crew of the first manned flight, Gemini 3, on March 23, 1965. Two and a half minutes after launch, the Gemini-Titan rocket was fifty miles high, traveling 6,700 miles per hour, with its two first-stage engines gulping 1,600 pounds of propellant every second, had already shed over three-quarters of its launch weight. The crew was being pressed into their ejection seats by nearly six times the force of gravity when suddenly the acceleration dropped off as the first-stage engines shutdown.

"Staging is really something," said John Young. "It's called 'fire in the hole' because you fire the second-stage engine before you get rid of the first stage. The firing blew out everything and fire came all around the vehicle and you could see it. That was a surprise to me. But it's only momentary, and with the second stage firing you get right out of there."

Gemini 3 splashed down after three orbits.

Three months later, Edward White became the first American to leave a space capsule and perform extravehicular activity, EVA, during Gemini 4's four days in space, which included sixty-two orbits.

Gemini 5 astronaut Gordon Cooper said it was a smoother ride than the liftoff on his previous Mercury flights, which had been launched by Atlas rockets. "The Titan had a thicker skin than the Atlas. The Atlas was literally a gas bag. Its skin was so thin it had to be pressurized to hold its shape. I remember going up the gantry for my Mercury flight and seeing the skin of the Atlas flexing. When you take that flexible skin and you launch it you get more oscillations, kind of like being on the end of a Slinky."

Gemini 5 demonstrated the eight-day endurance necessary for an Apollo lunar mission with the first use of fuel cells to generate its electrical power.

Gemini 6 was aborted. Wally Schirra and Tom Stafford heard the pre-valves open, the turbopumps whir, and the Titan begin to rumble with the power of 430,000 pounds of thrust. All indications were that Gemini 6 had lifted off from Cape Canaveral. Then the engines shut down.

"A light came on in the spacecraft saying we had liftoff," Schirra said. "I heard from the blockhouse that the clock had started, which means we had lifted off. But I knew we had not lifted off. It was a gut feeling. Stafford didn't know what was going on, but I had the experience of a Mercury flight, and my butt told me we hadn't left the pad."

Schirra's butt was right. A small electrical connector had vibrated loose a split-second before launch, sending a spurious shutdown signal to the Titan's first-stage engines. Had the Titan actually left the pad before engine cutoff, the missile would have toppled over, engulfing Schirra and Stafford in flames. If so, their only chance at survival would have been the Gemini's untried ejection seats.

Gemini 6A and 7 accomplished the first space rendezvous in December 1965, and Gemini 7 set a fourteen-day endurance record. Rendezvous in orbit is a complex maneuver as the forces on a spacecraft are much different than on an airplane.[5] We needed this experience for the lunar flights.

Gemini 8, 9 and 10 performed various rendezvous and docking maneuvers, including a simulation of lunar module rendezvous, plus numerous scientific, technological and medical experiments. EVA time was 2 hours, 9 minutes, a new EVA endurance record. During EVA maneuvers, Cernan's visor became fogged, and he was unable to test the Air Force astronaut maneuvering unit.

Charles Conrad and Richard Gordon, in Gemini 11, set the highest altitude record for a manned non-lunar mission of 854 miles in September 1966.[6]

The final Gemini spacecraft, Gemini 12, was launched in November 1966. Like Schirra and Stafford, Buzz Aldrin and Jim Lovell knew when it left the pad. After the mission, Aldrin said, "There was no doubt in Jim Lovell's and my mind when we lifted off. Our bodies knew it. We could feel the acceleration straight off." He proved that useful work could be accomplished outside a spacecraft without life-threatening exhaustion.

We learned much from Project Gemini. We were now ready to proceed further with Project Apollo and putting men on the Moon.

First "Space Walk." Ed White, June 1965.

Chapter 22 - The Spirit of Apollo

I have learned to use the word impossible with the greatest caution. - Wernher von Braun

Huntsville, 1961

I HAD A REPUTATION as an ultraconservative engineer. I liked to rely on the safe and proven, to move ahead only in small, measured steps, and to test, test, and test again down to every component of every subsystem. A good example of this was the two-stage Saturn I. It was first flown with a live first stage using Rocketdyne H-1 engines[1] and a dummy second stage. This was repeated three more times before a live second stage,[2] with Pratt & Whitney RL-10 engines, was added. Six vehicles of this configuration were launched. A total of ten Saturn I rockets were launched between October 1961 and July 1965, with missions carrying preliminary versions of the Apollo spacecraft.

The next small step was the Saturn IB with an uprated first stage and a new second stage, with a single Rocketdyne J-2 engine.[3] These rockets were used in the early phases of Apollo.

Development on the Saturn V officially began in January 1961, when it was announced we were to build what was then called the C-5 rocket. It was a follow-up rocket to the successful Jupiter series, and was eventually called the Saturn V. Saturn V would be the launch vehicle to land a man on the Moon. It was a three-stage rocket with the first stage powered by five Rocketdyne F-1 engines,[4] the second stage by five J-2 engines and the third stage by one J-2 engine.

<><><>

ALL OF THE SATURN WORK, in one way or another, tied in with our plans to land a man on the Moon. Initially, though, I had a different vehicle in mind. In late 1958, I published in *This Week* magazine, a novella, *First Men to the Moon*, where I described what NASA later called a "direct ascent" approach to a lunar landing. This consisted of a large rocket that would go directly to the Moon, and, as it turned around on approaching the Moon, would send a landing stage to the surface. The stage, upon leaving the Moon, would propel itself back to Earth.

This concept proved far too aggressive. It required an extremely large rocket, which NASA called Nova. The first stage of Nova would be, by some estimates, fifty feet in diameter, 400 feet tall and require a liftoff thrust of twelve million pounds.

My advanced missions' team and I favored a more practical design, which NASA called Earth-orbit rendezvous, or simply EOR. In this approach, we would have the lunar lander assembled and fueled in Earth orbit, then proceed to the Moon, land there and return to Earth. EOR became the baseline.

However, a young aeronautical engineer from the Langley Center preempted EOR. John Houbolt did some studies, which showed it would require far less propellants to have a stage circle the Moon with the lander going to the surface, returning to the stage then returning to Earth. His concept, the lunar-orbit rendezvous, LOR, avoided having the entire vehicle descend to and leave the lunar surface, which would require more propellants.

Initially, Houbolt's LOR met with considerable opposition, because it was considered to have more risk. No space rendezvous had been performed at that time, and a missed rendezvous in lunar orbit meant the astronauts couldn't return to Earth. However, an Earth rendezvous, if missed, still allowed the astronauts to return to Earth in the lunar lander. Later, the risk of a LOR became less meaningful after Gemini 6A and 7's successful rendezvous in Earth orbit.

The "mode decision" between EOR and LOR went on for months. Huntsville was concentrating on the launch vehicle and Houston on its spacecraft. Finally, on June 7, 1962, after a long day of meetings in Huntsville with Houston people present, I got up and said, "Well, gentlemen, I have listened to the arguments; I'm proud of the work you have done. Now I'll tell you the position of the center. Here are MSFC's[5] preferences in order: 1) LOR, preferably with the development of an unmanned, fully automatic one-way C-5 lunar logistics vehicle, 2) EOR, 3) C-5 Direct, and 4) Nova Direct.

"I prefer LOR, because it has the highest confidence factor of successful accomplishment within this decade. Also, it has an adequate performance margin, in that it uses conventional, storable propellants in the spacecraft and the ability to separate the reentry vehicle from the lunar landing vehicle. On the organizational side, LOR has the cleanest managerial interfaces, that is, there would be fewer meetings and coordination work than with the other approaches."

I paused, then continued, "John Houbolt of Langley was the first to advocate LOR. Against this background, it can, therefore, be concluded that the issue of 'invented here' versus 'not invented here' does not apply to either us at Marshall or to the Manned Spacecraft Center at Houston. Both centers have embraced a scheme suggested by a third source."

In addition, EOR required two launch vehicles, one to put the spacecraft in Earth orbit and a tanker to fuel the spacecraft for the lunar mission. This made EOR expensive and complex. An alternative: sending one vehicle to Earth orbit with enough fuel for the spacecraft made the launch vehicle and spacecraft extremely large. Landing a large 90-foot spacecraft on the Moon was a challenge, too.

Thus, the mode decision was made, and LOR became the baseline for the Apollo program. However, it was later decided that LOR would include a manned vehicle. The boosters for the program were the Saturn IB for the early Apollo Earth orbit flights, and the Saturn V for the later lunar flights.

<><><>

APOLLO HAD A THREE-PART SPACECRAFT: the first part was the Command Module, which was the crew's quarters and flight control section; the second part was the Service Module, for the propulsion and spacecraft support systems; the two modules, when together, were called the Command Service Module, CSM; and third part was the Lunar Module, to take two of the crew to the lunar surface, support them on the Moon, and return them to the CSM in lunar orbit.

During a preflight test, on January 27, 1967, for what was to be the first manned Apollo mission, later called Apollo 1, a fire claimed the lives of astronauts Gus Grissom, Ed White, and Roger Chaffee. It was an event I'll never forget. I was at a dinner party at the International Club in Washington with Gemini and Apollo corporate leaders when Lee Atwood,[6] got a phone call. He became very ashen-faced and turned to Bob Gilruth and said, "Bob, we've had a tragedy."

Lee and several others went to the Cape, and I was left at a very depressing party with the corporate leaders. Several of them got drunk because they were very upset. The tragedy caused a massive investigation, a reorganization of the entire Apollo program and it delayed the first manned launch until late 1968. NASA was guilt-ridden for not having seen the fire hazard from using high-pressure oxygen on the pad since the Mercury program.

<><><>

THE SATURN V ROCKET was 363 feet tall, about the height of a 36-story-tall building, and 60 feet taller than the Statue of Liberty. Fully fueled for liftoff, the Saturn V weighed 6.2 million pounds and could lift about fifty tons of payload into low Earth orbit. It was first launched in November 1967 as Apollo 4[7] and again in April 1968 as Apollo 6. Both launches were without crews but proved the rocket for manned launches. The first stage lifted the rocket to an altitude of about forty-two miles. The second stage carried it from there almost into orbit. The third stage placed the Apollo spacecraft into Earth orbit and pushed it toward the Moon. The first two stages fell into the ocean after separation. The third stage either stayed in space or hit the Moon.

Apollo 5 was the first unmanned flight of the Lunar Module, which would later carry astronauts to the Moon's surface. It lifted off on January 22, 1968, with a Saturn IB rocket on an Earth-orbit flight.

Apollo 7, launched on October 11, 1968, was the first U.S. three-man mission. The crew orbited the Earth 163 times in ten days and twenty hours. It was a successful test of the Command Service Module. It was the only manned Apollo mission launched by a Saturn IB rocket, the remaining missions by a Saturn V.

Apollo 8, launched two months later, was the first mission to take a crew around the Moon and was a prelude to landing humans on the Moon. The crew of Apollo 8 sent the first live televised pictures of the Earth and the Moon back to Earth, and read from the creation story in the Book of Genesis, on Christmas Eve 1968, "In the beginning God created the heaven and the Earth…"

The Apollo 9 mission was the first manned flight of all lunar hardware in Earth orbit and the first manned flight of the Lunar Module. The pilot, Russel Schweickart, performed a thirty-seven-minute Extra Vehicular Activity. Human reactions to space and weightlessness were tested in 152 orbits.

The Apollo 10 mission was a complete staging of the Apollo 11 mission without actually landing on the Moon. It was the first mission to travel to the Moon with the entire Apollo spacecraft configuration. Astronauts Thomas Stafford and Eugene Cernan descended inside the Lunar Module to within 50,000 feet of the lunar surface achieving the closest approach to the Moon before Apollo 11, which was launched two months later.

On July 19, 1969, we launched Apollo 11. The next day, Mike Collins remained in the Command Module while Neil Armstrong and Buzz Aldrin took the Lunar Module to the lunar surface. Millions of people watched on TV as Neil Armstrong put his foot on the Moon and said, "That's one small step for man, one giant leap for mankind." Armstrong and Aldrin returned to the Command Module and returned to Earth on July 22nd.

I had travelled to Mission Control at Houston to watch the historic landing, then returned to Huntsville. There, the community had a celebration and hero's welcome for me at Courthouse Square. I was carried on the shoulders of four members of the city council to the rostrum, where I said, "To be borne by a rocket to the Moon is impressive, but to be borne up these steps by admirers is almost as impressive!" I emphasized, "The space flights must continue. The ultimate destiny of man is no longer confined to Earth." I talked some about the remaining lunar missions, and I concluded with, "Maybe one of these days we'll even have a man on Mars."

Apollo 12 through 17 followed. All were successful except for Apollo 13. Commander James Lovell, just after a TV broadcast, said, "This is the crew of Apollo 13 wishing everybody there a nice evening, and we're just about ready to close out our inspection of Aquarius and get back for a pleasant evening in Odyssey. Good night."

Nine minutes later, oxygen tank No. 2 exploded, causing the No. 1 tank to also fail. The Command Module's normal supply of electricity, light and water was lost, and they were about 200,000 miles from Earth.

The event was a sharp bang and vibration at 9:08 p.m. April 13, 1970. Jack Swigert, Command Module Pilot, saw a warning light that accompanied the bang and said, "Houston, we've had a problem here." Instead of landing on the Moon, they circled it. Then, through the efforts of a very capable crew and mission control in Houston, Apollo 13 reverted to a free-return trajectory[8] and safely returned to Earth.

The six missions that landed on the Moon returned a wealth of scientific data and almost 880 pounds of lunar samples. Scientific instruments were used to retrieve data on the Moon's magnetic fields, geological characteristics, and exposure to meteoroids and solar wind to name a few.

Gene Cernan, the last man on the Moon, in December 1972, transmitted back to Earth, "I'm on the surface; and, as I take man's last step from the surface, back home for some time to come - but we believe not too long into the future - I'd like to just say what I believe history will record. That America's challenge of today has forged man's destiny of tomorrow. And, as we leave the Moon at Taurus-Littrow, we leave as we came and, God willing, as we shall return, with peace and hope for all mankind. Godspeed, the crew of Apollo 17."

In addition, there were many spinoffs from the whole space program that were useful here on Earth. There were advancements made in medicine, computing, materials, and food preservation to name just a few things that came out of the Space Race.

<><><>

IN SEPTEMBER 1969, the Space Task Force, headed up by Vice President Spiro Agnew, produced a report, "The Post-Apollo Space Program: Directions for the Future." It defined a program based on Saturn-Apollo, which included a manned landing on Mars. I whole-heartedly supported this report, as did Tom Paine, NASA Administrator, and George Mueller, Associate Administrator of NASA's Office of Manned Space Flight.

Surprisingly, later in September, Paine asked me to transfer to Washington and become head of planning for NASA. I would develop and sell this twenty-year program to the President, Congress and the nation. This was to be kept confidential until Paine could test the waters on this idea.

I had mixed thoughts and feelings on this proposal. It would certainly be an opportunity to develop space exploration and utilization beyond the Moon landings. On the personal side, Maria would once again enjoy big city life as she had growing up in Germany, but I would have to leave my team, many on the team had worked with me for twenty years.

In January 1970, while I was on vacation in Florida, the plan was leaked to the press. Paine had wanted a news conference to make the announcement with the President's approval backing it. With things messed up, Paine wanted me in Washington. However, I hadn't left with specific plans where in Florida I would be. Paine had the Florida Highway Patrol look for me, and they finally found me at a hotel in Palm Beach Shores.

There was an official press release in my name. Then, I stood before my co-workers at Marshall, and told them, "I feel like a navy skipper leaving his ship – but not a sinking ship! I'm leaving Marshall with nostalgia. I have my heart in Marshall. I love this place. I helped build it up. I've spent ten years here doing what was urgent, and regrettably not doing what was essential."

And so it was. In March 1970, we moved to Washington.

We put together a comprehensive twenty-year plan consisting of three separate Skylab space stations in Earth orbit, a simple space shuttle, continued production of Saturns and other rockets for scientific, unmanned missions to Mars, the Sun and the planets. Later, there would be a larger shuttle, a larger and permanent space station, a permanent lunar base and a manned mission to Mars.

However, the plan was not accepted, by the Nixon administration. Budgetary problems and other issues were deemed more important, so instead of moving forward, the NASA budget was cut severely. Tom Paine left NASA and rejoined General Electric.

I was very disappointed. Maria supported me and encouraged me through this drastic change at NASA. As a result, in 1972, I decided to retire from NASA and work for Fairchild Industries of Germantown, Maryland as vice president of engineering and development. There, I could work on some space projects of particular importance.

I worked on satellites for the most part. In 1973, I was diagnosed with a kidney tumor. I thought with its removal and radiation therapy I was OK, but in 1975, I had to have a malignant growth in my intestines removed along with a section of the intestine. All through these difficult times, I continued to promote space technology. However, the cancer spread and I had to resign from Fairchild in January 1977.

Epilogue

WERHNER VON BRAUN developed pancreatic cancer in 1977 and died of the disease at Alexandria, Virginia on June 16, 1977. He was buried at Ivy Hill Cemetery in Alexandria. His simple tombstone read:

WERNHER VON BRAUN, 1912-1977
Psalms 19:1*

*The heavens declare the glory of God; and the firmament showeth His handiwork.

The crater, von Braun, on the Moon is named after him. He received twelve honorary doctorates, one from the Technical University of Berlin from which he had graduated. In 1976, he was inducted into the International Space Hall of Fame at the New Mexico Museum of Space History.

Von Braun was responsible for the creation of the Research Institute at the University of Alabama in Huntsville. As a result of his vision, the university is one of the leading universities in the nation for NASA-sponsored research. The building housing the university's Research Institute was named in his honor in 2000, Von Braun Research Hall. The Rocket City Astronomical Association, in Huntsville was founded by von Braun and was later renamed after him, the Von Braun Astronomical Society.

In addition, Wernher von Braun wrote several books: *The Mars*

Project, (1952), *The Exploration of Mars* (1956), *First Men to the Moon* (1960), *Saturn Rocket Systems and Space Exploration* (1965), *History of Rocketry and Space Travel* (1966), *Space Frontier* (1968), *The Rockets Red Glare* (1976), *Project Mars: A Technical Tale* (published in 2006).

Wernher von Braum, 1912-1977
(Photo from the Public Domain)

Part Six

Epilogue

AFTER THE FALL OF THE SOVIET UNION in 1991, Russia and other former Soviet countries found themselves in dire economic times. They were open to Western companies to create businesses to shore up their economies. Conversely, Western companies sought to do business with them if it could prove fruitful.

In 1995, that Lockheed-Martin and Pratt & Whitney marketing personnel visited NPO Energomash, a Russian rocket engine manufacturer located in Moscow. To the Americans' surprise, the Russian engines were more advanced by at least ten years than anything produced in the West. This resulted in a joint venture between the three companies. Pratt & Whitney would redesign the Russian Space Shuttle engine for the Lockheed-Martin Atlas launch vehicle, and NPO Energomash would manufacture it in Moscow.

As part of the proposal for this effort, the author visited Dnipropetrovsk, Ukraine, where a derivative of the Russian Space Shuttle engine, the RD-171, was being installed in the Russian launch vehicle, the Zenit. Technicians tested the engine's structural integrity, electrical continuity, and hydraulic system functionality before and after installation. There was no computer on the engine. All control of the engine was by relays and solenoids.

The P&W chief engineer worked with NPO Energomash to redesign the RD-171 engine for the Atlas launch vehicle, and the author developed the pre-launch checkout procedures and associated ground support equipment. The new engine was designated the RD-180, and the new launch vehicle became the Atlas III.

A tremendous amount of resources went into the development of the RD-170 for the Russian space shuttle, the Energia. We learned the Soviets had 80 development engines built prior to production. This resulted in a truly reusable RD-170 certified for nine firings. The SSME, touted as reusable, had to be rebuilt after every flight. Thus, the RD-170, and its derivatives, were more reliable, as well.

On the humorous side, the P&W chief engineer kidded the Energomash chief engineer that with all the test engines for the Energia, he single-handedly sunk the Soviet economy. He, in turn, commented they thought the US Space Shuttle was disinformation---a bad design to throw off the design and development of the Russian Space Shuttle. The Russians would not use solid boosters for a manned vehicle because they were less reliable than liquid propellant engines.

Secondly, they would not use liquid hydrogen for a booster engine because the higher energy of hydrogen fuel is offset by a severe weight penalty of the larger and heavier structure and storage tank because of the low density of the liquid hydrogen. Interestingly, the failure of the Challenger Space Shuttle was due to the solid boosters and the failure of the Columbia was due to the large liquid hydrogen external tank when the foam on the tank broke off during launch damaging tiles on the leading edge of the orbiter.

But there's more to the story. The conventional wisdom throughout the Space Race and beyond was the Soviets got their rocket engine technology from the Germans after World War II. They acquired a German V-2 engine, a "gas-generator" engine, but used very little of its design. Instead, they developed a more powerful "staged combustion" engine, in the 1950s which were used on several launch vehicles from that era. The American SSME was the first U.S. staged combustion engine, which went into service in 1981 on the Space Shuttle Columbia.

Comparing the SSME with the RD-170. The SSME had a main combustion chamber pressure of about 3000 psi while the RD-170 had a pressure of about 3900 psi. Chamber pressure is a measure of performance. Following the Apollo program cancellation and U.S. funding cuts, the Soviets' continued high-level space program funding produced a more advanced space shuttle. Because of economic problems in the 1980s, the Soviets only launched Energia twice before cancelling the program.

The RD-170 engine's continued as the RD-171 for the Russian Zenit launch vehicle and the RD-180 for the Atlas III, which was first launched from Cape Canaveral Air Force Station on 24 May 2000. Engineers developed the Atlas III from the highly successful Atlas II, which had two stages. The first stage was a modified Atlas II with the RD-180 engine, and the upper stage retained the Centaur powered by the highly successful P&W RL-10 engine. After six launches, the Atlas III was replaced with the Atlas V, which had a modified structure enabling it to take the full thrust capability of the RD-180.

The engine has flown successfully on all six Atlas III flights and on all 99 Atlas V flights to date. However, the RD-180 is being phased out because it is being produced in Russia.

The Legacy

There are no constraints on the human mind, no walls around the human spirit, no barriers to our progress except those we ourselves erect. – Ronald Reagan

THE APOLLO PROGRAM has been called the greatest technological achievement in human history. It stimulated many areas of technology. For example -

The prototype inertial guidance system for the program could not be relied on completely and would need to be manually realigned during the flight. To assist the astronauts in this task, and to help them control the fly-by-wire systems in their new Apollo spacecraft, a small, lightweight computer was proposed by MIT. In the early 1960s, computers still tended to take up entire rooms. To miniaturize the computer enough to pack it into a modestly sized spacecraft, new technology was needed. The designers turned to an emerging invention called the integrated circuit. Fairchild Semiconductor was one of the few companies experimenting with these new micro-electronic components at the time and helped improve the performance of these miniature circuits for the space program.

Thus, the new micro-electronic technologies made affordable by Apollo led to the pocket calculators of the 1970s, the simple home computers and the internet of the 1980s, the world wide web of the 1990s, and the video streams and social networks of the 21st century. Today's cell phones, wireless equipment, iPads, etc. are a result of space technology development.

The design of the launch vehicle prompted improvements in bonding dissimilar materials, machining exotic alloys and new and improved approaches in manufacturing and production.

Biomedical and physiological monitoring systems, developed for manned spaceflight, are now used in hospitals today.

Exercise equipment, memory foam used in some mattresses, dialysis technology and freeze-dried food (in particular food sterilization and package sealing techniques), stay-dry clothing, and even no-fog ski goggles have their roots in space science.

Today, over a thousand artificial satellites orbit the Earth, relaying communications data around the planet and facilitating remote sensing of weather, vegetation, and human movements for the nations who employ them.

All of this derives from research initially driven by the Space Race.

Bibliography

Articles:

Mikhail S. Arlazorov, "Konstantin Eduardovich Tsiolkovsky, Soviet Scientist," *Encyclopedia Britannica.*

Robert H. Goddard, "A Method of Reaching Extreme Altitudes," *Smithsonian Miscellaneous Collections*, Volume 71, Number 2, City of Washington, Smithsonian Institution, 1919.

Stan Griffin, "A Very Early Space Pioneer."

"The Apollo Program," National Air and Space Museum files.

Books:

Arthur I. Berman, *The Physical Principles of Astronautics*, (New York and London, John Wiley and Sons, Inc. 1962).

Matthew A. Bille and Erika R. Lishock, *The First Space Race: Launching the World's First Satellites*, (College Station, Texas, Texas A&M University Press, 2004)

David A. Clary, *Rocket Man: Robert H. Goddard and the Birth of the Space Age*, (New York, Hyperion, 2003)

Jamie Doran and Piers Bizony, *Starman: The Truth behind the Legend of Yuri Gagarin,* (New York, Walker and Company, 2011)

Chris Gainor, *To a Distant Day: The Rocket Pioneers* (Omaha, University of Nebraska Press, 2013)

James Harford, *Korolev: How One Man Masterminded the Soviet Drive to Beat America to the Moon*, (New York, John Wiley & Sons, Inc., 1997)

A. Kosmodemyansky, *Konstantin E. Tsiolkovsky, His Life and Work*, (Honolulu, Hawaii, University Press of the Pacific, 2000, reprinted from the 1956 edition)

Michael J. Neufeld, *von Braun: Dreamer of Space, Engineer of War*, (New York, Vintage Books, a division of Random House, Inc., 2007)

Compiled by V.N. Sokolsky, *Selected Works of Konstantin E. Tsiolkovsky*, (Honolulu, Hawaii, University Press of the Pacific, 2004, reprinted from the 1968 edition)

Asif A. Siddiqi, *Sputnik and the Soviet Space Challenge,* (Gainesville, Florida, University Press of Florida, 2003)

Tom Streissguth, *Rocket Man, The Story of Robert Goddard*, (Minneapolis, Carolrhoda Books, Inc., 1995)

George P. Sutton, *Rocket Propulsion Elements*, (New York and London, John Wiley and Sons, 1963)

Helen B. Walters, *Hermann Oberth: Father of Space Travel*, (New York, The Macmillian Company, 1962)

Bob Ward, *Dr. Space: The Life of Wernher von Braun*, (Annapolis, Maryland, Naval Institute Press, 2005)

Websites:

"Esther," Goddard Memorial Association, goddardmemorial.org.

"Konstantin E. Tsiolkovsky," September 22, 2010, nasa.gov.

"Korolev," "RD-100," "RD-101, et al" *Encyclopedia Astronautica,* www.astronautix.com/astros/korolev.htm

Cliff Lethbridge, "Konstantin Eduardovitch Tsiolkovsky," spaceline.org.

"The Secrets of Tsiolkovsky," http://www.thelivingmoon.com

About the Author

JOSEPH F ALCOCK is a former aerospace engineer. He was employed for more than forty years by Pratt & Whitney Rocketdyne (PWR), a division of United Technologies and was a project engineer on many of the company's commercial and military jet engines and rocket engines. Before he retired, he was the pre-launch operations manager for the booster engines in the Atlas III and Atlas V launch vehicles. He has written many proposals for PWR and several articles, stories and instruction manuals for local organizations since he retired in 2006.

End Notes

Prologue

[1] Alexis de Tocqueville, a French aristocrat and historian traveled widely in America and took extensive notes about his observations and reflections. He is famous for the following quotes regarding America:

- "The American Republic will endure until the day Congress discovers that it can bribe the public with the public's money."
- "America is great because she is good. If America ceases to be good, America will cease to be great."
- "Liberty cannot be established without morality, nor morality without faith."

Chapter 1

[1] Boris Pasternak's father.

[2] It is a story of the Baltimore Gun Club, a post- American Civil War society of weapons enthusiasts, and their attempts to build an enormous space gun to launch three people in a projectile with the goal of landing on the moon.

[3] Mendeleyev is generally considered the co-creator of the Periodic Table of Elements and predicted several elements, which had not yet been discovered.

[4] The concept was reinvented several times in the 20th century but was introduced to the general public in 1979 by the publication of Arthur C. Clarke's novel, *The Fountains of Paradise*, in which engineers construct a space elevator on top of a mountain peak in the fictional island country of Taprobane.

[5] In 2000, an American scientist, Bradley C. Edwards, suggested creating a 62,000-mile-long paper-thin ribbon using a carbon composite material. Supported by the NASA Institute for Advanced Concepts, Edwards' work was expanded to research the design, deployment and operation of a space elevator.

Chapter 2

[1] Reactive device, or reactive apparatus, is a term derived from Newton's Second Law, which states that for every action, there is an equal and opposite reaction. In the case of a rocket, exhaust gases create a force in the opposite direction causing the rocket to move in that direction.

The difference between Newton's equation and Tsiolkovsky's equation is Newton's was for a body with a fixed mass and Tsiolkovsky's was for a body with a variable mass caused by the depletion of the propellants during flight.

[2] 8,000 meters per second is 28,800 kilometers per hour or 18,000 miles per hour. 8,000 meters per second is the *ideal* velocity required to reach a circular Earth orbit.

8,000 meters per second produces a centrifugal force, which is equal and opposite to the force of gravity. Thus, a satellite travelling at this speed remains in orbit around the Earth.

[3] The Rocket Equation calculates the change in flight velocity to equal the reactive device's exhaust velocity times the natural logarithm of 1 plus the ratio of the vehicle initial mass of the propellants-to-the mass of the vehicle without propellants. In mathematical terms:

$$V = Ve \times \ln (1 + M_2/M_1)$$

where V = the maximum velocity of the rocket, assuming the initial velocity is zero,
Ve = the velocity of the exhaust gases from the reactive device,
$\ln$ = the natural logarithm,
M_2 = initial mass of the propellants,
M_1 = initial mass of the vehicle, without propellants.

Note that the above equation is an idealized relationship. A more exact equation would take into account gravity effects especially at lift-off and air resistance to name just a few factors not included in the equation. Also, it assumes all the propellants are used. Computer programs are used today to handle the complexity resulting from these other factors in order to accurately calculate V.

Tsiolkovsky assumed an exhaust velocity, Ve, of 5700 meters per second as the absolute maximum exhaust velocity that could be attained, using liquid hydrogen and liquid oxygen. Then a M_2/M_1 ratio of 3 would yield the final vehicle velocity, V, of 7902 meters per second, about what is needed, theoretically, to put a satellite in Earth orbit.

[4] Later the same year, 1897, the Rocket Equation was published in the thesis of the Russian mathematician I. V. Meshchersky, *Dynamics of a Point of Variable Mass*, St. Petersburg, 1897.

[5] *Science Review.*

[6] The boiling point of hydrogen is -252.9 deg C (- 423.2 deg F),
oxygen is -183 deg C (-297.3 deg F).

Chapter 3

[1] This is referred to as the ideal escape velocity. It can be determined from equating the kinetic energy of an object to the work necessary to overcome gravity.

$\frac{1}{2} * m * Vesc^2 = m * g * R$ Eq. (1)

where m = mass of the object
Vesc = escape velocity
R = radial distance the object moves from the center of the Earth
g = force of gravity on the object

$R = Ro + h$ Eq. (2)
$g = g_o * (Ro/R)^2$ Eq. (3)

where, R_0 = radius of the Earth at sea level
h = object's height above the Earth
g_0 = force of gravity at sea level

rearranging Eq. (1) and substituting Eqs. (2) and (3) yields:
$Vesc = R_0 * SQRT(2 * g_0 / (R_0 + h))$

The ideal velocity to escape the Earth = 11,179 meters / sec, from the Moon = 2,347 meters / sec.

[2] Previously, in 1903, Tsiolkovsky assumed propellant weight of 3 times the empty vehicle weight (or 75 % of the total vehicle weight) was needed to reach Earth orbit. In 1911, it appears he factored in the effects of gravity when he raised the propellant weight to 80 percent of the total vehicle weight.

[3] 7,912 meters / sec is the ideal velocity for a satellite to go into a circular Earth orbit. This is from the requirement that the centrifugal force acting out toward space equals the gravitational force acting toward Earth when an object is in Earth orbit. See the drawing in Chapter 2. In mathematical terms:

$Fcent = Fgravity$
$m * Vorb^2 / R = m * g$ Eq. (1)

where, m = mass of the object
Vorb = orbital velocity
R = radial distance the object is from the center of the Earth
g = force of gravity on the object in orbit

$R = Ro + h$ Eq. (2)
$g = g_o * (Ro/R)^2$ Eq. (3)

where, R_0 = radius of the Earth at sea level

h = object's height above the Earth
g_0 = force of gravity at sea level

rearranging Eq. (1) and substituting Eqs. (2) and (3) yields:
$Vorb = R_0 * SQRT (g_0 / (R_0 + h))$

Note the escape velocity is greater than the orbital velocity by the square root of 2, which is 1.4142.

[4] Today, the acceleration effects during launch is usually called "g forces," expressed in multiples of the gravity force, g. The number of g's typically increase during a launch. A vehicle is typically throttled back, at some point, so the g's are tolerable to the vehicle structure and humans. Four g's would make a 200 lb. man "weigh" 800 lbs.

[5] From a lecture given by Sergei P. Korolev given on 17 September 1957 on the centennial of Tsiolkovsky's birth.

Chapter 4

[1] Named after William Congreve (1772-1828), a British artillery commander. His rockets initially used cardboard casings. However, by 1806, the rocket cases were made from sheet iron. The propulsion was from a mixture of the same ingredients as gunpowder, the mixture of which varied with the different sizes of rocket. The warheads had side-mounted brackets, which were used to attach wooden sticks of differing lengths according to the sizes of rocket. The sticks served to provide some guidance to the rocket. By 1813, the rockets were available in three classes: light, medium and heavy.

[2] Goddard calculated that only 2 percent of the chemical energy from the combustion process was converted into kinetic energy in the rocket exhaust gases. Therefore, he added a de Laval nozzle, from steam turbine designs, to achieve a higher exhaust kinetic energy.

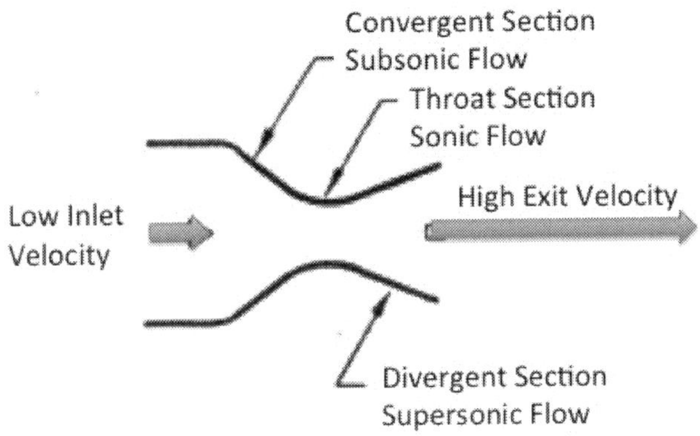

de Laval Nozzle

[3] Patent numbers 1,102,653 and 1,103,503.

[4] The device resembled a bazooka, which was used extensively in World War Two. However, Goddard never patented the device.

[5] Hun was a derogatory description of the Germans during the Great War, referring back to Attila the Hun.

Chapter 5

[1] After Sachs lit the igniter, made from match heads, he lit an alcohol tank under the LOX tank. When Goddard opened a valve, pressurized oxygen from the LOX tank entered the fuel system, where it raised the pressure there forcing both gasoline and oxygen to enter the combustion chamber and ignite.

[2] Multiple spray nozzles are used in almost all present day rocket engines.

[3] This method of cooling the chamber is still used today. This is a fuel-rich condition. Locally, there is excess fuel, which won't burn as all the oxygen has been used. The unburned fuel vaporizes, which reduces the temperature.

Chapter 6

[1] As president of Clark University, he ordered in 1922, the lights be turned off while Scott Nearing was addressing a Liberal Club on socialism on the campus

o, which won him great renown. He also banned *The Nation*, a leftist magazine, from the Clark University campus.

[2] About 300 pounds, according to the static tests.

[3] Von Karman was one of the famous aerospace engineers of the twentieth century. He is known for his work in supersonic and hypersonic aerodynamics.

[4] Von Karman later said that Goddard needed to move the center of gravity forward and improve the combustion stability to produce a more constant thrust. Sounding rockets to this day have no gyroscopic system for stability.

[5] This was an early type of jet engine.

Chapter 7

[1] Goddard tried a different method for cooling in 1923. It was a regeneratively cooled engine, which circulated liquid oxygen around the outside of the combustion chamber, but he thought the system was far too complicated and abandoned it. However, regenerative cooling is used in most liquid propellant rocket engines today.

[2] The addition of fuel after the stoichiometric fuel-air ratio (where all the available air has been combusted) is reached quenches combustion temperature.

[3] Turbopumps are used on most rockets today. The following schematics compare an engine with turbopumps with one with tank pressurization:

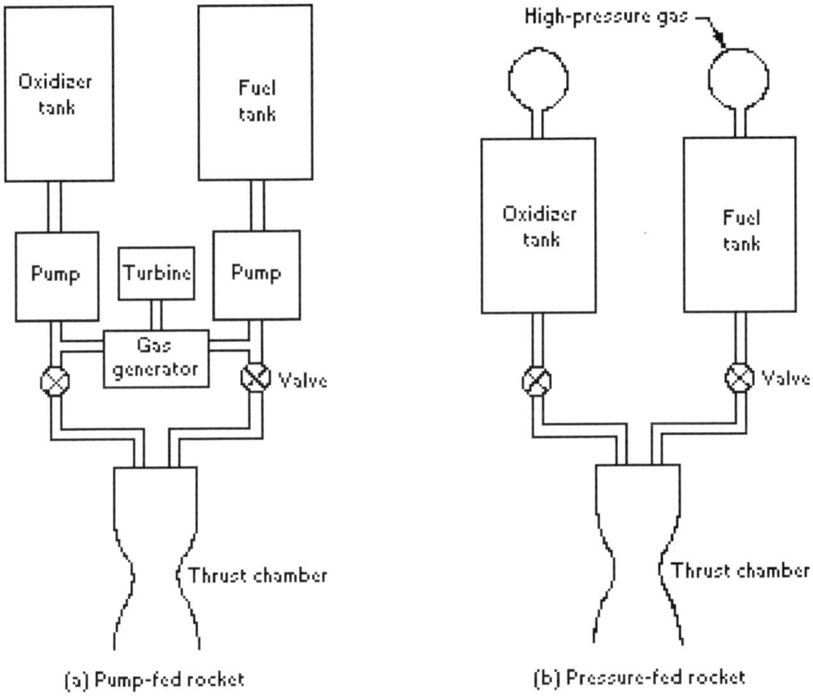

4 How much of Goddard's work was copied by the Germans was debated for some time after his death. Certainly, the basic design was similar, but the Germans had their own designs, which were more advanced.

5 This is a box built and used by American rocket pioneer Robert H. Goddard in experiments in 1916 to determine the amount of magnesium flash powder to be carried in a unmanned rocket to strike the surface of the Moon to signal its arrival.

Chapter 8

1 Today Sibiu, Romania

2 About $2500 US.

3 *Die Rakete zu den Planetenräumen.*

Chapter 9

[1] This principle was later adopted and is still in use today. The following is a schematic of regenerative cooling of the chamber and nozzle using fuel before it enters the chamber:

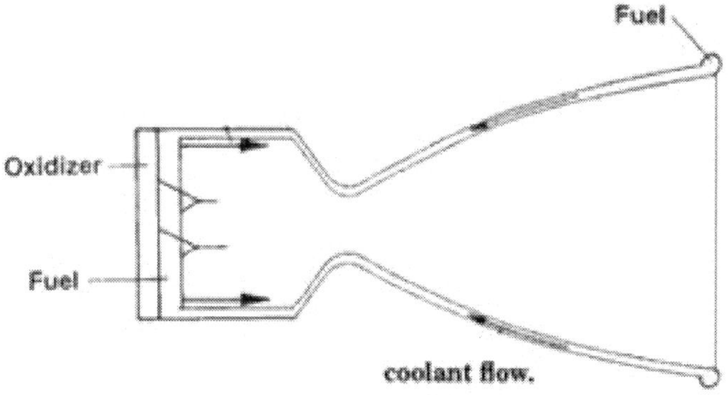

[2] NASA's Meteoroid Environment Office (MEO), located at NASA's Marshall Space Flight Center in Huntsville, Alabama, tracks meteoroids to understand the flux and the associated risk of meteoroids impacting spacecraft traveling in and beyond Earth's orbit. Meteoroids impacting spacecraft are a quantifiable risk as they can puncture pressurized volumes (i.e. space station modules, propellant tanks) or destroy components (i.e. engines, electronics). While meteoroids have never been definitively proven to have destroyed a spacecraft, there have been several in-flight anomalies attributed to meteoroid impacts, e.g. loss of satellite control during a meteor shower.

[3] A space mirror has been suggested to control climate change. However the size calculated, by a senior staff scientist at Lawrence Livermore Labs, of such a mirror is about 600,000 square miles.

[4] There was an article in Life magazine's July 23, 1945 issue titled "The German Space Mirror." The article credits unnamed U.S. Army experts for revealing a Nazi plan to construct the extraordinary device. Consisting of a reflective, slightly concave disk approximately one mile in diameter, the sun gun would focus solar rays onto enemy cities -- and burn them.

[5] The Defense Science Board Task Force wrote a paper in June 2001 on High Energy Laser Weapon Systems Application, which details the use of space mirrors to enhance space weapons.

[6] *Wege zur Raumschiffahart*

[7] *Frau im Mond*

[8] About $400.

[9] Government Institute for Chemistry and Technology.

[10] *Kegelduese*

[11] Chamber

[12] 15 pounds.

Chapter 10

[1] 900 kg thrust = about 1980 pounds thrust.

[2] 2204 pounds.

[3] 360 miles per hour.

[4] 45 feet, 11 inches and 27,600 pounds.

[5] 50,000 pounds.

[6] 55 miles.

[7] 3,600 miles per hour.

Chapter 11

[1] *Wasserfall* in German.

[2] *Westfalisch-Anhaltische Springstoff A.G.* (Westphalian-Anhaltian Explosives Co.)

[3] Menshen im Weltraum in German.

[4] October 24, 1954 issue, page 4.

[5] An asteroid is a small rocky body orbiting the sun. Large numbers of these, ranging in size from nearly 600 miles across (Ceres) to dust particles, are found in the asteroid belt between the orbits of Mars and Jupiter, though some have

more eccentric orbits, and a few pass close to the earth or enter the atmosphere as meteors.

[6] From *Man into Space.*

[7] This principle was used on all Atlas launch vehicles until the Atlas V, which was first launched in 2002.

[8] Water in alcohol is used to help cool the combustion chamber wall.

[9] *Stoff und Leben*

Chapter 12

[1] Reaktivni Nauchno-Isledovatelski Institut, or RNII.

[2] 30 December 1906 in the Old Russian calendar.

[3] Now Ukraine.

[4] Society of Aviation and Aerial Navigation of Ukraine and Crimea.

[5] MVTU is the Russian abbreviation.

[6] A ballistic missile is a projectile which moves exclusively under gravitational forces after its initial launching.

[7] A rocket, or aircraft, has six degrees of freedom (DOF). Korolev's auto-gyro designs controlled either two or three of the possible six DOFs. The six possible degrees of freedom are: three possible linear movements and three possible rotational movements. The sketch below illustrates these movements:

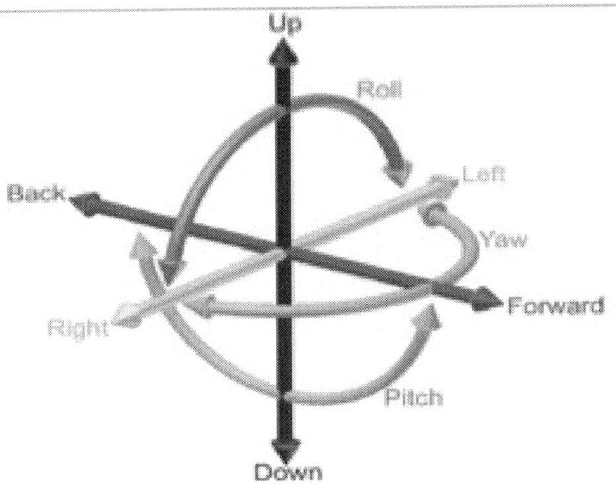

Sketch showing six degrees of freedom.
(From the Public Domain)

Chapter 13

[1] Located just north of the Sea of Japan.

[2] Located about 1400 kilometers southeast of Moscow.

[3] Located at the confluence of the Volga and Kazanka Rivers in European Russia.

[4] About 210 to 260 miles per hour.

Chapter 14

[1] Renamed Volgograd by Nikita Khrushchev in 1961 as part of his de-Stalinization program.

[2] Aluminum mixed with copper for added strength. It was widely used in aviation in the 1930s.

[3] The tank wall also served as the skin of the vehicle.

[4] A "dirty" bomb.

[5] RD-101 engine with 37,209 kg (82,033 lbs) thrust at sea level.

[6] The RD-100 engine, for the R-1, used liquid oxygen and alcohol (75% alcohol and 25% water). The RD-101 engine, for the R-2, used liquid oxygen and a higher alcohol concentration (96% alcohol and 4% water).

[7] Methyl alcohol, also known as "wood alcohol," because it is made from distilling wood is poisonous.

[8] The RD-110 engine designed for both the G-4 and the R-3 had a chamber pressure of 58 atmospheres, a thrust of 119,920 kg at sea level.

[9] This was the RD-102, which was replaced by the RD-103 and RD-103M engines.

[10] The "M" designated nuclear capability.

[11] The RD-103M engine.

[12] The R-4 was an air-to-air missile, and there was never a R-6 vehicle for some unknown reason.

[13] Four RD-107 engines and one RD-108 engine. Each engine had propellants from the turbopumps, which were supplied to small, hinged vernier engines for attitude control.

[14] The Soviet nuclear warhead was more primitive which dictated a large vehicle to deliver it.

[15] Russian for fellow traveler of Earth.

Chapter 15

[1] Located on the Baltic Sea.

[2] First Secretary of the Central Committee of the Communist Party of the Soviet Union at the time. In March 1958, Khrushchev became the premier of the Soviet Union.

[3] Laika died due to a failure of the thermal control system.

[4] Three of these failures were Soviet rockets. They failed twice due to longitudinal resonance of boosters, high vibration and once due to loss of lubrication to hydrogen peroxide pump, respectively.

[5] It missed due to a malfunction of the upper-stage separation system.

[6] An oxygen-rich environment was used to minimize weight.

[7] The external covering to increase streamlining and reduce drag while in the atmosphere.

[8] In the Lower Volga River region.

Chapter 16

[1] *Zenit* is zenith in Russian.

[2] *Molinya* is lightning.in Russian.

[3] *Soyez* is union in Russian.

[4] *Voskhod* is ascent or dawn in Russian.

[5] One of the most interesting facts about the Soviet/Russian engines was that they developed oxygen rich, staged combustion in the early 60s and that most all their engines since then have used this cycle.

Chapter 17

[1] About $1,700 in 1930.

[2] V*erein fur Raumschiffahrt*, Society for Spaceship Travel.

Chaptern18

[1] German Air Force.

[2] About $1.2 million.

[3] Heinrich Himmler, from 1943 on, was both Chief of German Police and Minister of the Interior, overseeing all internal and external police and security forces, including the *Gestapo* (Secret State Police).

[4] Abbreviation of *Geheime Staatspolizei*, or the Secret State Police was the official secret police of Nazi Germany and German-occupied Europe.

[5] Albert Speer was one of Hitler's inner circle and Minister of Armaments and War Production.

[6] In 1944, Himmler convinced Adolf Hitler to put the V-2 project directly under SS control. On 8 August 1944, Hans Kammler replaced Walter Dornberger as its director and on 31 January 1945, he became head of all missile projects.

Chapter 19

[1] VABV for *Vorhaben zur Besonderen Verwendung.*

[2] Fins on the rocket and vanes in the exhaust were used to control the vehicle's direction in flight.

[3] The WAC or WAC Corporal was the first sounding rocket developed in the U.S. Begun as a spinoff of the Corporal program, the WAC was a "little sister" to the larger Corporal. It was designed and built jointly by the Douglas Aircraft Company and the Guggenheim Aeronautical Laboratory. The WAC Corporal was a hypergolic liquid-propellant rocket. Fuming nitric acid was the oxidizer and a mixture of aniline and furfuryl alcohol (with the later addition of hydrazine) was the fuel.

[4] There was no major damage and the unexpected turn to the south was traced to a faulty gyroscope.

[5] Vehicles were not necessarily launched in the same order as their designations.

[6] The first V-2s, in Germany, used a simple analog computer that adjusted the direction of the rocket relative to north (azimuth). The flying distance was controlled by timing the engine cut-off as the flight path was nearly parabolic. A ground-based Doppler system, or later on-board integrating accelerometers, measured the flight speed and distance for shut-off purposes. Some later V-2s used radio signals transmitted from the ground, to keep the missile on course. More advanced V-2s were guided by a gyroscopic system that sent signals to aerodynamic steering tabs on the external fins and the vanes located in the engine exhaust.

Chapter 20

[1] The Tullahoma facility was the Arnold Engineering and Development Center, or AEDC, named after Henry "Hap" Arnold, the father of the U.S. Air Force.

[2] The Juno I was a Jupiter-C, a three-stage rocket, with an additional fourth stage.

- The first stage had a Rocketdyne A-7 engine which used liquid oxygen, as the oxidizer, and "Hydyne" (60% unsymmetrical, dimethylhydrazine and 40% diethylenetriamine), as fuel; propellant feed was by a turbopump driven by steam which was produced from 90% hydrogen peroxide decomposed by a catalyst bed to produce steam.

- The second stage had eleven Jet Propulsion Lab (JPL) scaled-down Sergeant rockets, which used polysulfide-aluminum and ammonium perchlorate as a solid propellant.
- The third stage had three JPL scaled-down Sergeant rockets.
- The fourth stage had one JPL scaled-down Sergeant rocket.

Chapter 21

[1] He also is credited with discovery of a new moon of Saturn in 1979, as well as radiation belts around that planet.

[2] *Explorer II* failed to reach orbit after launch on March 5, 1958 when the fourth stage failed to ignite.

[3] It used a Rocketdyne A-7 engine with 75% ethyl alcohol/25% water and liquid oxygen as propellants. The vehicle also had an automatic abort system.

[4] Hypergolics were storable propellants, ideal for an ICBM that required immediate launch capability.

[5] Should a spacecraft increase its speed to catch up with another, the result is that it goes into a higher and slower orbit and the distance thereby increases. The right procedure is to slow down and go to a lower orbit first, and then later to increase speed and go to the same orbit as the other. To practice these maneuvers special rendezvous and docking simulators were built for the astronauts.

[6] This is still the record today.

Chapter 22

[1] It had eight Rocketdyne H-1 LOX/RP-1 engines.

[2] It consisted of six Pratt & Whitney RL-10 LOX/liquid hydrogen engines.

[3] The J-2 was a LOX/hydrogen engine.

[4] The F-1 was a LOX/RP-1 engine producing about 1.5 million pounds of thrust.

[5] Marshall Space Flight Center.

[6] Chairman of the Board, North American Aviation.

---

[7] Because of a NASA change in the numbering system after Apollo 1, the next Apollo launch was Apollo 4.

[8] A free-return trajectory relies on the Moon's gravity as a steering device. If the spacecraft's trajectory comes in from behind the Moon, its velocity increases as it picks up a little of the Moon's momentum. This is sometimes called a slingshot effect.

Over the course of the Apollo flights, NASA initially used the free-return trajectory, because it provided the safest return. As confidence increased, modifications to the pure free-return trajectory were used, which provided more lunar landing sites. The early flights up to and including Apollo 11 used the pure free-return trajectory, which set the spacecraft directly back to Earth.

Apollo 12, 13 and 14 started as free return, but once they were safely on their way the flight path was modified to improve the approach to the Moon. Apollo 13, after the LOX tank exploded, had a short engine burn to restore the free-return trajectory for a safer return to Earth. After Apollo 14, a non-free-return trajectory was used, but again these flights could employ a short burn for a safe return.

Made in United States
Orlando, FL
04 November 2026

90203667R00129